D1386384

AS / Year 1
Physics

Exam Board: AQA

Revising for Physics exams is stressful, that's for sure — even just getting your notes sorted out can leave you needing a lie down. But help is at hand...

This brilliant CGP book explains **everything you'll need to learn** (and nothing you won't), all in a straightforward style that's easy to get your head around. We've also included **exam questions** to test how ready you are for the real thing.

There's even a free Online Edition you can read on your computer or tablet!

How to get your free Online Edition

Go to **cgpbooks.co.uk/extras** and enter this code...

0541 4661 7075 8482

This code only works for one person. If somebody else has used this book before you, they might have already claimed the Online Edition.

A-Level revision? It has to be CGP!

GREENWICH LIBRARIES
3 8028 02234207 2

Published by CGP

Editors:
David Maliphant, Rachael Marshall, Matteo Orsini Jones, Sam Pilgrim, Frances Rooney,
Charlotte Whiteley and Sarah Williams

Contributors:
Tony Alldridge, Jane Cartwright, Peter Cecil, Barbara Mascetti, John Myers and Andy Williams

GREENWICH LIBRARIES WE	
3 8028 02234207 2	
Askews & Holts	16-Oct-2015
530	£10.99
4807999	

With thanks to Mark Edwards, Ian Francis and Glenn Rogers for the proofreading.

ISBN: 978 1 78294292 4

www.cgpbooks.co.uk

Clipart from Corel®
Printed by Elanders Ltd, Newcastle upon Tyne.

Based on the classic CGP style created by Richard Parsons.

Text, design, layout and original illustrations © Coordination Group Publications Ltd. (CGP) 2015
All rights reserved.

Photocopying more than one chapter of this book is not permitted. Extra copies are available from CGP.
0870 750 1242 • www.cgpbooks.co.uk

Contents

Section 1 — Particles

Atomic Structure 2
Stable and Unstable Nuclei 4
Particles and Antiparticles 6
Forces and Exchange Particles 8
Classification of Particles 10
Quarks 13

Section 2 — Electromagnetic Radiation and Quantum Phenomena

The Photoelectric Effect 16
Energy Levels and Photon Emission 18
Wave-Particle Duality 20

Section 3 — Waves

Progressive Waves 22
Longitudinal and Transverse Waves 24
Superposition and Coherence 26
Stationary Waves 28
Diffraction 30
Two-Source Interference 32
Diffraction Gratings 34
Refractive Index 36

Section 4 — Mechanics

Scalars and Vectors 38
Forces 40
Moments 42
Mass, Weight and Centre of Mass 44
Displacement-Time Graphs 46
Velocity-Time and Acceleration-Time Graphs 48
Motion With Uniform Acceleration 50
Acceleration Due to Gravity 52
Projectile Motion 54
Newton's Laws of Motion 56
Drag, Lift and Terminal Speed 58
Momentum and Impulse 60
Work and Power 62
Conservation of Energy and Efficiency 64

Section 5 — Materials

Properties of Materials 66
Stress and Strain 68
The Young Modulus 70
Stress-Strain and Force-Extension Graphs 72

Section 6 — Electricity

Current, Potential Difference and Resistance 74
I/V Characteristics 76
Resistivity and Superconductivity 78
Electrical Energy and Power 80
E.m.f. and Internal Resistance 82
Conservation of Energy and Charge 84
The Potential Divider 86

Practical and Investigative Skills

Experiment Design 88
Uncertainty and Errors 90
Presenting and Evaluating Data 92

Answers 94
Index 100

Atomic Structure

"So what did you do today, Johnny?" "Particle physics, Mum." "How nice dear — done with times tables then?" Yeah, well, it's not exactly the ***easiest*** *topic in the world, but it's a darn sight more interesting than biology.*

Atoms *are made up of* ***Protons, Neutrons*** *and* ***Electrons***

Inside **every atom**, there's a **nucleus** containing **protons** and **neutrons**.
Protons and **neutrons** are both known as **nucleons**. **Orbiting** this core are the **electrons**.
This is the **nuclear model** of the atom.

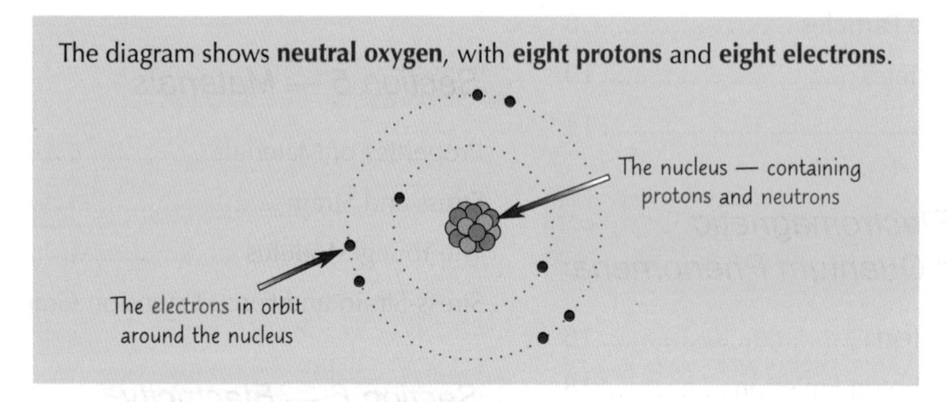

The diagram shows **neutral oxygen**, with **eight protons** and **eight electrons**.

Tom dreamt of becoming a nuclear model when he grew up.

The particles in an atom have different **properties**. Their charges and masses are so **tiny** that it's often easier to talk about their **relative charge** and **relative mass**.

You need to learn the values in the orange columns — you won't be given them in the exam.

Particle	Charge (coulombs, C)	Mass (kg)	Relative Charge	Relative Mass
Proton	$+1.60 \times 10^{-19}$	1.67×10^{-27}	+1	1
Neutron	0	1.67×10^{-27}	0	1
Electron	-1.60×10^{-19}	9.11×10^{-31}	−1	0.0005

The ***Proton Number*** *is the* ***Number*** *of* ***Protons*** *in the Nucleus*

No... really.

The **proton number** is sometimes called the **atomic number**, and has the **symbol Z** (I'm sure it makes sense to someone). **Z** is just the **number of protons** in the nucleus.

It's the **proton number** that **defines** the **element** — **no two elements** will have the **same** number of protons.

In a **neutral atom**, the number of **electrons equals** the number of **protons**.
The element's **reactions** and **chemical behaviour** depend on the number of **electrons**.
So the **proton number** tells you a lot about its **chemical properties**.

A particle with a different number of electrons to protons is called an ion.

The ***Nucleon Number*** *is the* ***Total Number*** *of* ***Protons*** *and* ***Neutrons***

The **nucleon number** is also called the **mass number**, and has the **symbol A** (*shrug*).
It tells you how many **protons** and **neutrons** are in the nucleus. Since each **proton or neutron** has a relative **mass** of (approximately) **1** and the electrons weigh virtually nothing, the **number** of **nucleons** is the same as the **atom's relative mass**.

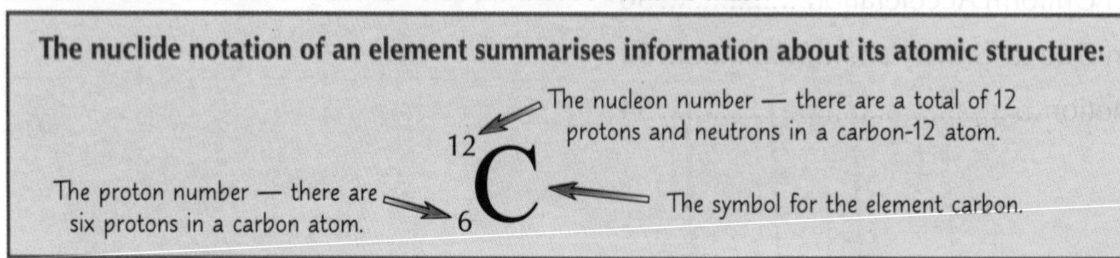

Atomic Structure

Isotopes have the *Same Proton Number*, but *Different Nucleon Numbers*

Atoms with the **same number of protons** but **different numbers of neutrons** are called **isotopes**.

Example: Hydrogen has three natural isotopes — hydrogen, deuterium and tritium.
Hydrogen has 1 proton and 0 neutrons.
Deuterium has 1 proton and 1 neutron.
Tritium has 1 proton and 2 neutrons.

Changing the number of **neutrons doesn't affect** the atom's **chemical** properties.

The **number of neutrons** affects the **stability** of the nucleus though.

Unstable nuclei may be **radioactive** and **decay** over time into different nuclei that are more stable (see p.5).

Radioactive Isotopes Can be Used to Find Out *How Old* Stuff Is

1) All living things contain the same percentage of radioactive **carbon-14** taken in from the atmosphere.
2) After they die, the amount of carbon-14 inside them **decreases** over time as it **decays** to stable elements.
3) Scientists can calculate the **approximate age** of archaeological finds made from dead **organic matter** (e.g. wood, bone) by using the **isotopic data** (amount of each isotope present) to find the percentage of **radioactive carbon-14** that's **left in** the object.

The *Specific Charge* of a Particle is Equal to its *Charge Over its Mass*

The **specific charge** of a particle is the ratio of its charge to its mass, given in coulombs per kilogram (C kg^{-1}). To calculate specific charge, you just divide the charge in C by the mass in kg.

$$\textbf{Specific charge} = \frac{\textbf{charge}}{\textbf{mass}}$$

You could be asked to find the specific charge of any particle, from a **fundamental particle** like an electron, to the nucleus of an atom or an ion.

A fundamental particle is one that you can't break up into anything smaller.

Example: Calculate the specific charge of a proton.

A proton has a **charge** of $+1.60 \times 10^{-19}$ C and a **mass** of 1.67×10^{-27} kg (see p.2).
So specific charge = $(+1.60 \times 10^{-19}) \div (1.67 \times 10^{-27}) = 9.580... \times 10^{7}$ = **9.58×10^{7} Ckg^{-1} (to 3 s.f.)**

In calculations, always give your answer to the smallest number of significant figures used in the question.

Practice Questions

Q1 List the particles that make up the atom and give their relative charges and relative masses.

Q2 Define the proton number and nucleon number.

Q3 Explain how the amount of carbon-14 in dead organic matter can tell scientists how old it is.

Q4 How could you calculate the specific charge of a particle?

Exam Questions

Q1 Describe the nuclear model of the atom. [2 marks]

Q2 Write down the numbers of protons, neutrons and electrons in a neutral atom of oxygen, $^{16}_{8}O$. [2 marks]

Q3 a) State what is meant by an 'isotope'. [1 mark]

b) State the similarities and differences between the properties of two isotopes of the same element. [2 marks]

Q4 An alpha particle is the nucleus of a $^{4}_{2}He$ atom. Calculate the specific charge of an alpha particle. [4 marks]

"Proton no. = no. of protons" — hardly nuclear physics is it... oh wait...

Physics is the science of all things great and small — on these pages you saw the small (like, really really small). That's why it's useful to know about things like relative charge and relative mass — 'one point six seven times ten to the power of negative twenty-seven' is a bit more of a mouthful than 'one'. It all makes perfect sense.

Stable and Unstable Nuclei

Keeping the nucleus stable requires a lot of effort — a bit like Physics then...

The **Strong Nuclear Force** Binds Nucleons Together

There are several different **forces** acting on the nucleons in a nucleus. The two you already know about are **electrostatic** forces from the protons' electric charges, and **gravitational** forces due to the masses of the particles.

If you do the calculations (don't worry, *you* don't have to) you find the repulsion from the **electrostatic force** is much, much **bigger** than the **gravitational** attraction. If these were the only forces acting in the nucleus, the nucleons would **fly apart**. So there must be **another attractive force** that **holds the nucleus together** — called the **strong nuclear force**. (The gravitational force is so small, you can just ignore it.)

The **strong nuclear force** is quite **complicated**:

1) To **hold the nucleus together**, it must be an **attractive force** that's **stronger** than the electrostatic force.
2) Experiments have shown that the strong nuclear force has a **very short range**. It can only hold nucleons together when they're separated by up to **a few femtometres** (1 fm = 1×10^{-15} m) — the size of a nucleus.
3) The **strength** of the strong nuclear force **quickly falls** beyond this distance (see the graph below).
4) Experiments also show that the strong nuclear force **works equally between all nucleons**. This means that the size of the force is the same whether it's proton-proton, neutron-neutron or proton-neutron.
5) At **very small separations**, the strong nuclear force must be **repulsive** or it would **crush** the nucleus to a **point**.

The **Size** of the Strong Nuclear Force **Varies** with **Nucleon Separation**

The **strong nuclear force** can be plotted on a **graph** to show how it changes with the **distance of separation** between **nucleons**. If the **electrostatic force** is also plotted, you can see the **relationship** between these **two forces**.

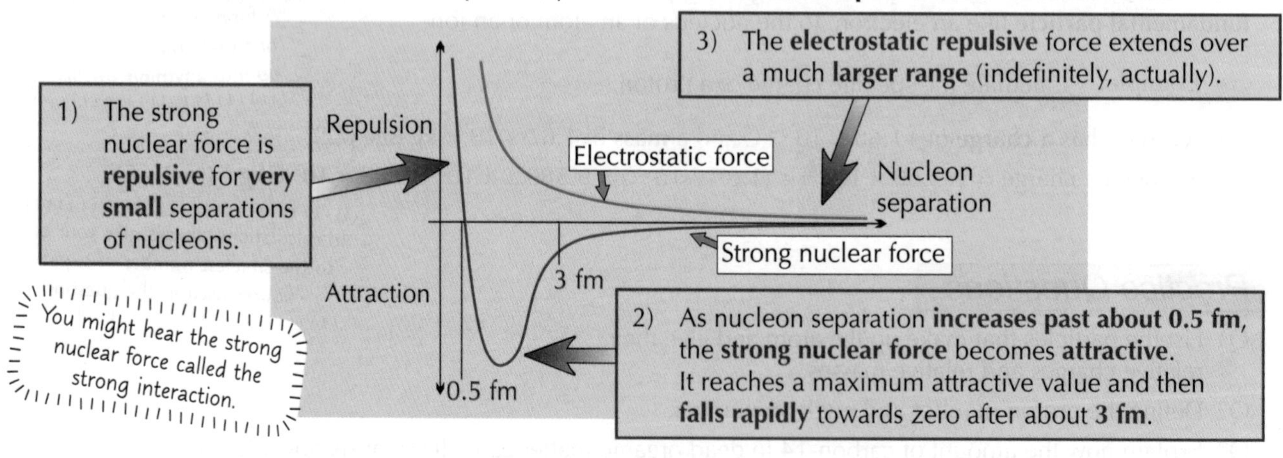

α **Emission** Happens in **Very Big Nuclei**

1) **Alpha emission** only happens in **very big** nuclei, like **uranium** and **radium**.
2) The **nuclei** of these atoms are just **too massive** for the strong nuclear force to keep them stable.
3) When an alpha particle is **emitted**:

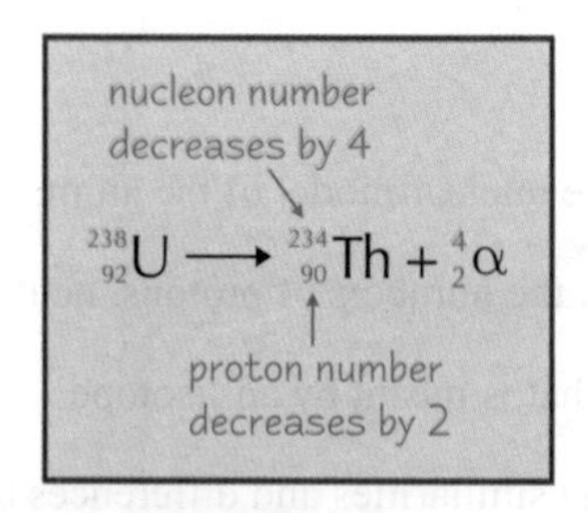

The **proton number decreases** by **two**, and the **nucleon number decreases** by **four**.

Alpha particles have a very **short range** — only a few cm in air. This can be seen by observing the tracks left by alpha particles in a **cloud chamber**. You could also use a **Geiger counter** (a device that measures the amount of ionising radiation). Bring it up close to the alpha source, then **move it away** slowly and observe how the **count rate drops**.

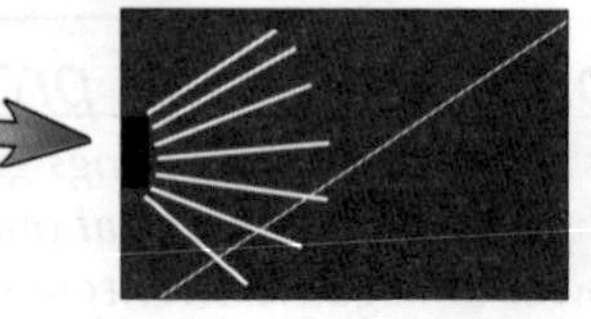

The thin line is a cosmic ray particle

Stable and Unstable Nuclei

β^- Emission Happens in Neutron-Rich Nuclei

1) **Beta-minus** (usually just called beta) decay is the emission of an **electron** from the **nucleus** along with an **antineutrino**.
2) Beta decay happens in isotopes that are unstable due to being **'neutron rich'** (i.e. they have too many more **neutrons** than **protons** in their nucleus).
3) When a nucleus ejects a beta particle, one of the **neutrons** in the nucleus is **changed** into a **proton**.

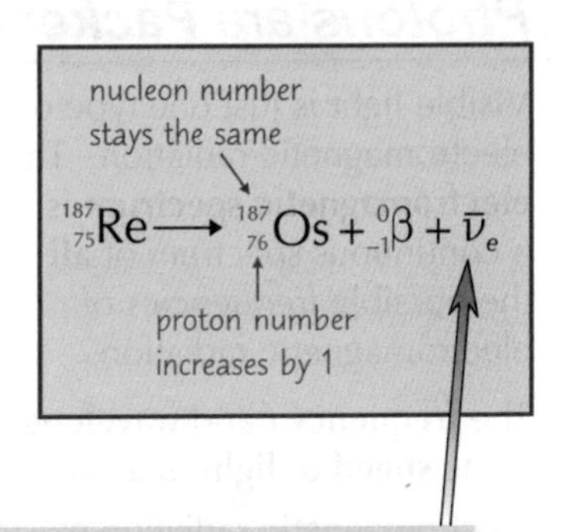

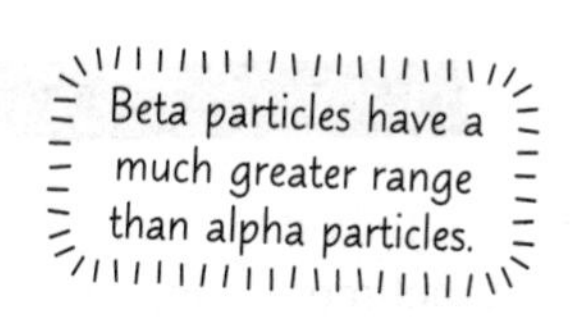

The **proton number increases** by **one**, and the **nucleon number stays the same.**

In beta decay, you get a **tiny neutral particle** called an **antineutrino** released. This antineutrino carries away some **energy** and **momentum**.

Neutrinos Were First Hypothesised Due to Observations of Beta Decay

1) Scientists originally thought that the **only** particle emitted from the nucleus during beta decay was an **electron**.
2) However, observations showed that the **energy** of the particles **after** the **beta decay** was **less** than it was **before**, which didn't fit with the principle of **conservation of energy** (p. 64).
3) In 1930 Wolfgang Pauli suggested **another particle** was being emitted too, and it carried away the **missing energy**. This particle had to be **neutral** (or charge wouldn't be **conserved** in beta decay) and had to have **zero** or **almost zero** mass (as it had never been **detected**).
4) Other discoveries led to Pauli's theory becoming accepted and the particle was named the **neutrino**. (We now know this particle was an antineutrino — p. 6).
5) The neutrino was eventually observed 25 years later, providing evidence for Pauli's hypothesis.

Practice Questions

Q1 What causes an electrostatic force inside the nucleus?

Q2 What evidence suggests the existence of a strong nuclear force?

Q3 Is the strong interaction attractive or repulsive at a nucleon separation of 2 fm?

Q4 Describe the changes that happen in the nucleus during alpha and beta-minus decay.

Q5 What observations led to the hypothesis of the existence of the neutrino?

Exam Questions

Q1 The strong nuclear force binds the nucleus together.

a) Explain why the force must be repulsive at very short distances. [1 mark]

b) Explain why a nucleus containing two protons in unstable, but one containing two protons and two neutrons is stable. [2 marks]

Q2 Radium-226 and potassium-40 are both unstable isotopes.

a) Radium-226 undergoes alpha decay to radon. Complete the balanced nuclear equation for this reaction:

$^{226}_{88}Ra \longrightarrow \quad Rn +$ [3 marks]

b) Potassium-40 (Z = 19, A = 40) undergoes beta decay to calcium. Write a balanced nuclear equation for this reaction. [4 marks]

The strong interaction's like nuclear glue...

Energy, momentum, charge and nucleon number (and several other things that you'll find out about in this section) are conserved in every nuclear reaction. That's why the antineutrino in beta decay has to be there.

Particles and Antiparticles

"I cannae do it Cap'n — their electron-antineutrino ray gun's interfering with my antineutron positron reading..."

Photons are Packets of Electromagnetic Radiation

Visible light is just one type of electromagnetic radiation. The **electromagnetic spectrum** is a continuous spectrum of **all** the possible frequencies of electromagnetic radiation.

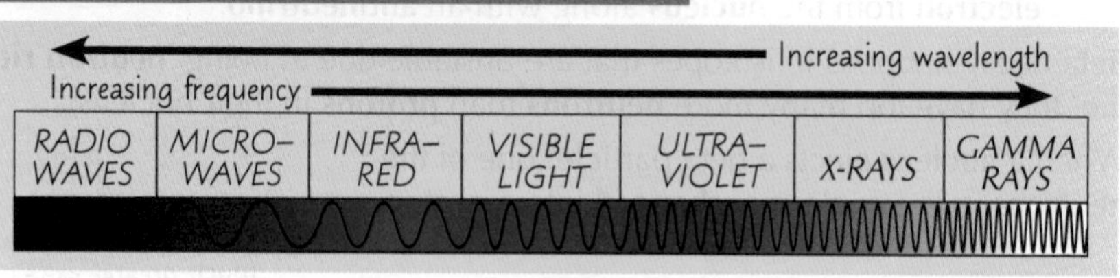

The **frequency** f and **wavelength** λ are linked by $f = \frac{c}{\lambda}$, where $c = 3.00 \times 10^8$ ms^{-1} is the **speed of light** in a vacuum (sometimes called the speed of light **in vacuo**).

Electromagnetic radiation exists as **photons** of energy (page 16). The **energy** of a photon depends on the frequency of the radiation: $E = hf = \frac{hc}{\lambda}$

h is the Planck constant, equal to 6.63×10^{-34} Js.

Every Particle has an Antiparticle

1) Each particle has a **matching antiparticle** with the **same mass** and **rest energy** (more later), but with **opposite charge** (amongst other things).
2) For instance, an **antiproton** is a **negatively-charged** particle with the same mass as the **proton**, and the **antineutrino** is the antiparticle of the **neutrino** — it doesn't do much either.

Particle/Antiparticle	Symbol	Relative Charge	Mass (kg)	Rest Energy (MeV)
proton	p	+1	$1.67(3) \times 10^{-27}$	938(.3)
antiproton	$\bar{p}$	–1		
neutron	n	0	$1.67(5) \times 10^{-27}$	939(.6)
antineutron	$\bar{n}$			
electron	e^-	–1	9.11×10^{-31}	0.51(1)
positron	e^+	+1		
neutrino	ν_e	0	0	0
antineutrino	$\bar{\nu}_e$			

These are actually an electron-neutrino and an electron-antineutrino (p. 12).

Luckily, in the exam you'll be given all the **masses** in kg and **rest energies** in MeV of each of these particles and their antiparticles. You just need to remember that the **mass** and **rest energy** are the **same** for a particle and its antiparticle. Neutrinos and antineutrinos are incredibly tiny — you can assume they have zero mass and zero rest energy.

1 MeV = 1×10^6 eV. There's more on eV (electron volts) on p. 18.

You can Create Matter and Antimatter from Energy

You've probably heard about the **equivalence** of energy and mass. It all comes out of Einstein's Special Theory of Relativity. **Energy** can turn into **mass** and **mass** can turn into **energy** if you know how. The **rest energy** of a particle is just the 'energy equivalent' of the particle's **mass**, measured in MeV. You can work it all out using the formula $E = mc^2$, but you won't be expected to do the calculations for AS.

When **energy** is converted into **mass** you get **equal amounts** of **matter** and **antimatter**.

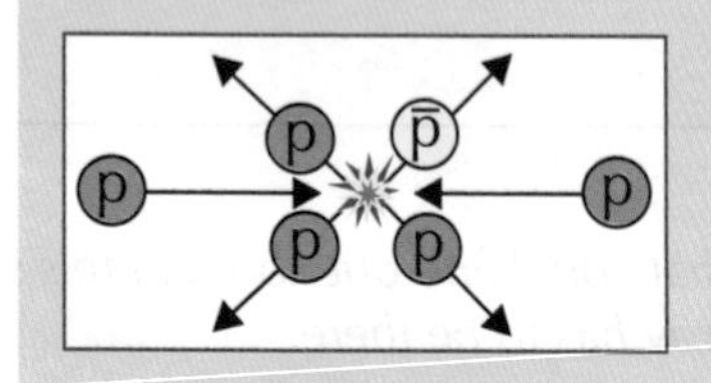

Fire **two protons** at each other at high speed and you'll end up with a lot of **energy** at the point of impact. This energy might be converted into **more particles.**

If an extra **proton** is formed then there will always be an **antiproton** to go with it. It's called **pair production.**

Particles and Antiparticles

Each Particle-Antiparticle Pair is Produced from a Single Photon

Energy that gets **converted** into **matter** and **antimatter** is in the form of a **photon** (p.16). Pair production only happens if **one photon** has enough energy to produce that much mass — only **gamma ray** photons have enough energy. It also tends to happen near a **nucleus**, which helps conserve momentum.

You usually get **electron-positron** pairs produced (rather than any other pair) — because they have a relatively **low mass**.

The particle tracks are curved because there's usually a magnetic field present in particle physics experiments. They curve in opposite directions because of the opposite charges on the electron and positron.

The **minimum energy** for a photon to undergo **pair production** is the **total rest energy** of the particles produced.
The particle and antiparticle each have a rest energy of E_0, so:

$$E_{min} = hf_{min} = 2E_0$$

The Opposite of Pair-Production is Annihilation

When a **particle** meets its **antiparticle** the result is **annihilation**. All the **mass** of the particle and antiparticle gets converted back to **energy**. Antiparticles can usually only exist for a fraction of a second before this happens, so you don't get them in ordinary matter.

OR

The electron and positron annihilate and their mass is converted into the energy of a pair of gamma ray photons to conserve momentum.

An annihilation is between a particle-antiparticle pair, which both have a rest energy E_0. **Both** photons need to have a **minimum energy**, E_{min}, which when added together equals at least $2E_0$ for **energy** to be **conserved** in this interaction. So $2E_{min} = 2E_0$ and:

$$E_{min} = hf_{min} = E_0$$

Example: Calculate the maximum wavelength of one of the photons produced when an electron and positron annihilate each other.

For annihilation, minimum photon energy $E_{min} = hf_{min} = E_0$. Remember $f = \frac{c}{\lambda}$, so $\frac{hc}{\lambda_{max}} = E_0$.

So $\lambda_{max} = \frac{hc}{E_0} = \frac{(6.63 \times 10^{-34}) \times (3.00 \times 10^{8})}{(0.511 \times 10^{6}) \times (1.60 \times 10^{-19})} = 2.432... \times 10^{-12} = \mathbf{2.43 \times 10^{-12}}$ **m (to 3 s.f.)**

The Planck constant is in J, so you need to convert E_0 from MeV to J.

Practice Questions

Q1 Describe the properties of an electron-antineutrino.

Q2 Give one similarity and one difference between a proton and an antiproton.

Q3 What is pair production?

Q4 What happens when a proton collides with an antiproton?

Exam Questions

Q1 Write down an equation for the reaction between a positron and an electron and state the name for this type of reaction. [2 marks]

Q2 Explain what causes extra particles to be created when two particles collide. [2 marks]

Q3 Give a reason why the reaction: $p + p \rightarrow p + p + n$ is not possible. [1 mark]

Q4 A photon produces an electron-positron pair, each with 9.84×10^{-14} J of energy. Calculate the frequency of the photon. [2 marks]

This really is Physics at its ~~hardest~~ grooviest...

Inertial dampers are off-line Captain.........oops, no — it's just these false ears making me feel dizzy. Anyway — you'd need to carry an awful lot of antimatter to provide enough energy to run a spaceship. Plus, it's not the easiest to store...

Forces and Exchange Particles

*Having learnt about all those lovely particles and antiparticles, you now have the esteemed privilege of learning about yet another weirdy thing called a **gauge boson**. To the casual observer this might not seem **entirely fair**. And I have to say, I'd be with them.*

Forces are Caused by *Particle Exchange*

You can't have **instantaneous action at a distance** (according to Einstein, anyway). So, when two particles **interact**, something must **happen** to let one particle know that the other one's there. That's the idea behind **exchange particles**.

1) **Repulsion** — Each time the **ball** is **thrown or caught** the people get **pushed apart**. It happens because the ball carries **momentum**.

Particle exchange also explains **attraction**, but you need a bit more imagination.

2) **Attraction** — Each time the **boomerang** is **thrown or caught** the people **get pushed together**. (In real life, you'd probably fall in first.)

These exchange particles are called **gauge bosons**.

The **repulsion** between two **protons** is caused by the **exchange** of **virtual photons**, which are the gauge bosons of the **electromagnetic** force. Gauge bosons are **virtual** particles — they only exist for a **very short time**.

There are *Four Fundamental Forces*

All forces in nature are caused by four **fundamental** forces — the strong nuclear force, the weak nuclear force, the electromagnetic force and gravity. Each one has its **own gauge boson** and these are the ones you have to learn:

Type of Interaction	Gauge Boson	Particles Affected
electromagnetic	virtual photon (symbol, γ)	charged particles only
weak	W^+, W^-	all types
strong	pions (π^+, π^-, π^0)	hadrons only

Particle physicists never **bother** about **gravity** because it's so incredibly **feeble** compared with the other types of interaction. Gravity only really **matters** when you've got **big masses** like **stars and planets**.

In the **strong nuclear force**, pions are described as being exchanged between **nucleons**. You might also see it described as **gluons** being exchanged between **quarks** (p. 13).

The *Larger* the *Mass* of the *Gauge Boson*, the *Shorter* the *Range* of the *Force*

1) The **W bosons** have a **mass** of about **100 times that of a proton**, which gives the weak force a **very short range**. Creating a **virtual W particle** uses **so much energy** that it can only exist for a **very short time** and it **can't travel far**.
2) On the other hand, the **photon** has **zero mass**, which gives you a force with **infinite range**.

You can use *Diagrams* to Show What's *Going In* and What's *Coming Out*

Particle interactions can be hard to get your head around. A **neat way** of **solving problems** is by **drawing simple diagrams** of particle interactions rather than doing **calculations**.

1) **Gauge bosons** are represented by **wiggly lines** (technical term).
2) Other **particles** are represented by **straight lines**.

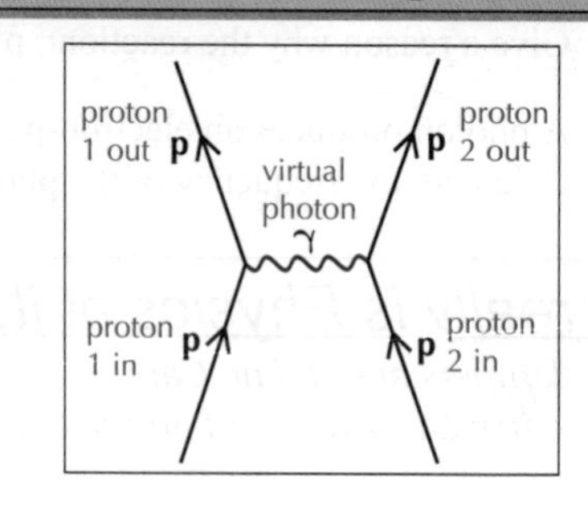

Forces and Exchange Particles

*You can draw simple diagrams of **loads** of interactions, but you **only** need to learn these ones for your exams.*

You Need to Be Able to **Draw Diagrams** of these **Interactions**

Electromagnetic Repulsion

This is the easiest of the lot. When two particles with **equal charges** get close to each other, they **repel**.

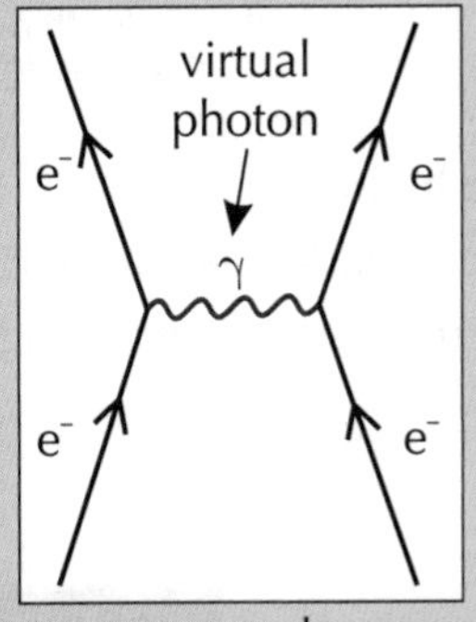

two electrons repelling each other

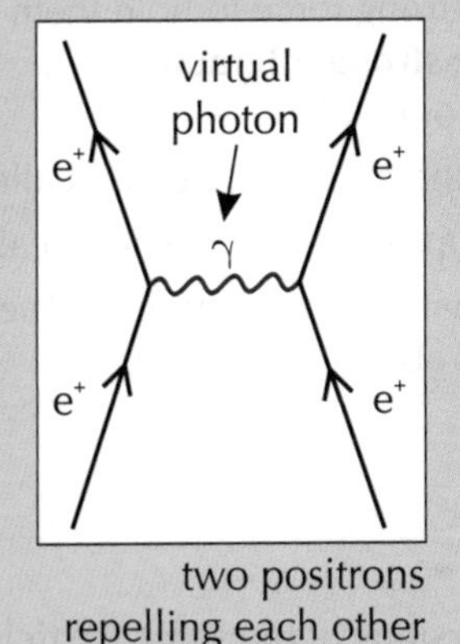

two positrons repelling each other

RULES FOR DRAWING PARTICLE INTERACTION DIAGRAMS:

1) **Incoming** particles start at the bottom of the diagram and move upwards.
2) The **baryons** (p.10) and **leptons** (p.12) can't cross from one side to the other.
3) Make sure the charges on both sides balance. The **W** bosons carry **charge** from one side of the diagram to the other.
4) A W^- particle going to the **left** has the same effect as a W^+ particle going to the **right**.

Electron Capture and Electron-proton Collisions

Electrons and protons are of course attracted by the **electromagnetic interaction** between them, but if a proton **captures** an electron, the **weak interaction** can make this interaction happen.

$$p + e^- \rightarrow n + \nu_e$$

electron capture

You also need to know about **electron-proton collisions**, where an electron **collides** with a proton. The equation is just the same as electron capture but in the diagram a **W^- boson** goes from the **electron** to the **proton** instead of a W^+ travelling the other way.

Beta-plus and Beta-minus Decay

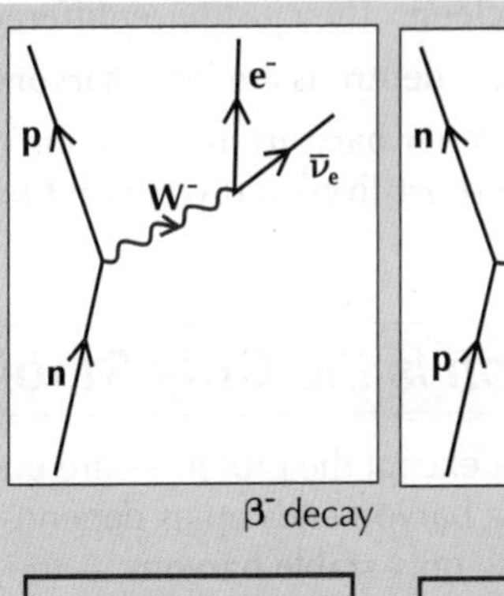

β⁻ decay

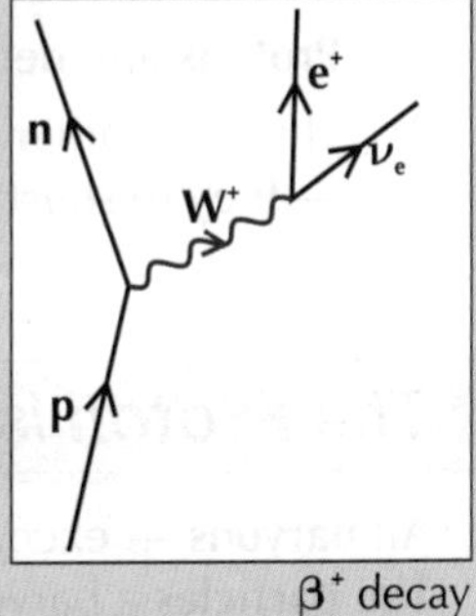

β⁺ decay

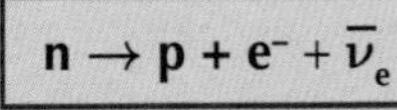

$$n \rightarrow p + e^- + \bar{\nu}_e$$

$$p \rightarrow n + e^+ + \nu_e$$

You get an **antineutrino** in β^- decay and a **neutrino** in β^+ decay so that **lepton number** (p.12) is conserved.

You'll see on p.14 that when a proton changes to a neutron or vice versa, it has to be the weak interaction.

Practice Questions

Q1 List the four fundamental forces in nature.

Q2 Explain what a virtual particle is.

Q3 Draw a simple diagram to show the particles involved in a beta-minus decay interaction.

Q4 Which gauge bosons are exchanged in weak interactions?

Exam Questions

Q1 Describe how the force of electromagnetic repulsion between two protons is explained by particle exchange. [2 marks]

Q2 Draw a diagram to show the particle interaction when an electron and a proton collide. Label all the particles involved and state clearly which type of interaction is involved. [3 marks]

I need a drink...

Urrrgghhhh... eyes... glazed... brain... melting... ears... bleeding... help me... help me...

help me...

Classification of Particles

There are loads of different types of particle apart from the ones you get in normal matter (protons, neutrons, etc.). They only appear in cosmic rays and in particle accelerators, and they often decay very quickly so they're difficult to get a handle on. Nonetheless, you need to learn about a load of them and their properties. Stick with it — you'll get there.

Hadrons are Particles that Feel the Strong Nuclear Force (e.g. Protons and Neutrons)

1) The **nucleus** of an atom is made up of **protons** and **neutrons** (déjà vu).
2) Since the **protons** are **positively charged** they need a strong force to hold them together. This is called the **strong nuclear force** or the **strong interaction** (who said physicists lack imagination...). See page 4 for details.
3) **Not all particles** can **feel** the **strong nuclear force** — the ones that **can** are called **hadrons**.
4) Hadrons aren't **fundamental** particles. They're made up of **smaller particles** called **quarks** (see pages 13–15).
5) There are **two** types of **hadrons** — **baryons** (and anti-baryons) and **mesons**. They're classified according to the number of **quarks** that make them up, but don't worry about that for now.

(Leptons are an example of particles that can't. See page 12.)

Protons and Neutrons are Baryons

1) It's helpful to think of **protons** and **neutrons** as **two versions** of the **same particle** — the **nucleon**. They just have **different electric charges**.
2) **Protons** and **neutrons** are both **baryons**.
3) There are **other baryons** that you don't get in normal matter — like **sigmas** (Σ) — they're **short-lived** and you **don't** need to **know about them** (woohoo!).

Baryon and Meson felt the strong interaction.

The Proton is the Only Stable Baryon

All **baryons** — except the proton — are **unstable**. This means that they **decay** to become other **particles**. The **particles** a baryon ends up as depends on what it started as, but it **always** includes a **proton**. **Protons** are the only **stable baryons** — they don't decay (as far as we know).

All baryons except protons decay to a **proton**.

Some theories predict that protons should decay with a very long half-life, but there's no evidence for it at the moment.

Antiprotons and Antineutrons are Antibaryons

The **antiparticles** of protons and neutrons — **antiprotons** and **antineutrons** — are **antibaryons**. But, if you remember from page 7, **antiparticles** are **annihilated** when they meet the corresponding **particle** — which means that you **don't** find **antibaryons** in ordinary matter.

The Number of Baryons in an Interaction is called the Baryon Number

The **baryon number** is the number of baryons. (A bit like **nucleon number** but including unusual baryons like Σ too.) The **proton** and the **neutron** each have a baryon number **B = +1**. **Antibaryons** have a baryon number **B = –1**. **Other particles** (i.e. things that aren't baryons) are given a baryon number **B = 0**.

Baryon number is a **quantum number** that must be **conserved** in any interaction — that means it can only take on a **certain set of values** (so you can't have 2.7981 baryons, or 1.991112 baryons... you get the idea).

When an **interaction** happens, the **baryon number** on either side of the interaction has to be the **same**. You can use this fact to **predict** whether an **interaction** will **happen** — if the numbers don't match, it can't happen.

The **total baryon number** in **any** particle interaction **never changes**.

Classification of Particles

Neutrons are Baryons that Decay into Protons

You saw on pages 5 and 9 that **beta decay** involves a **neutron** changing into a **proton**. This happens when there are many **more neutrons** than **protons** in a nucleus or when a **neutron** is **by itself**, **outside** of a nucleus. **Beta decay** is caused by the **weak interaction** (see page 14).

When a neutron decays, it forms a **proton**, an **electron** and an **antineutrino**:

$$n \rightarrow p + e^- + \bar{\nu}_e$$

Electrons and **antineutrinos** aren't baryons (they're **leptons**, as you'll see on the next page), so they have a baryon number **B = 0**. **Neutrons** and **protons** are baryons, so have a baryon number **B = 1**. This means that the **baryon numbers** on both sides are **equal** (to 1), so the interaction **can** happen.

The Mesons You Need to Know About are Pions and Kaons

The second type of hadron you need to know about is the **meson**.

1) **All mesons** are **unstable** and have **baryon number B = 0** (because they're not baryons).
2) **Pions** (π-mesons) are the **lightest mesons**. You get **three versions** with different **electric charges** — π^+, π^0 and π^-. You get **loads** of pions in **high-energy particle collisions** like those studied at the **CERN** particle accelerator.
3) **Kaons** (**K**-mesons) are **heavier** and more **unstable** than **pions**. You get different ones like K^+ and K^0. Kaons have a very **short lifetime** and **decay** into **pions**.
4) Pions and kaons were **discovered** in **cosmic rays** — cosmic ray showers are a source of both particles. You can observe the tracks of these particles with a **cloud chamber** (see p.4).
5) Mesons **interact** with **baryons** via the **strong force**.

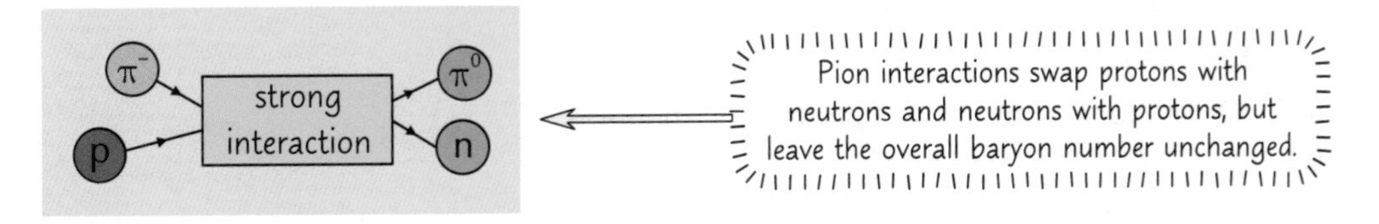

Summary of Hadron Properties

DON'T PANIC if you don't understand all this yet. For now, just **learn** these properties. You'll need to work through to the end of page 15 to see how it **all fits together**.

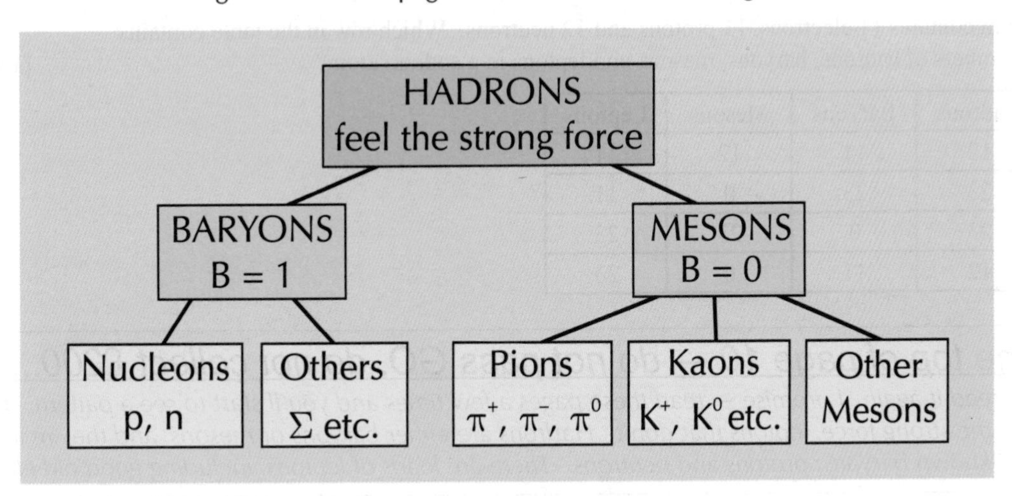

Classification of Particles

Leptons *(e.g. Electrons and Neutrinos)* **Don't** *Feel the* **Strong Nuclear Force**

1) **Leptons** are **fundamental particles** and they **don't** feel the **strong nuclear force**. They only really **interact** with other particles via the **weak interaction** (along with a bit of gravitational force and the electromagnetic force as well if they're charged).
2) **Electrons** (e^-) are **stable** and very **familiar**, but — you guessed it — there are also **other leptons**, such as the **muon** (μ^-), that are just like **heavy electrons**.
3) **Muons** are **unstable**, and **decay** eventually into **ordinary electrons**.
4) The **electron** and **muon** leptons each come with their **own neutrino**, ν_e and ν_μ.
5) **Neutrinos** have **zero** or **almost zero mass**, and **zero electric charge** — so they don't do much. **Neutrinos** only take part in **weak interactions** (see p.14). In fact, a neutrino can **pass right through the Earth** without **anything** happening to it.

You Have to **Count** *the* **Types** *of Lepton* **Separately**

Like the baryon number, the **lepton number** is just the number of **leptons**. **Each lepton** is given a **lepton number** of **+1**, but the **electron** and **muon** types of lepton have to be **counted separately**. You get **different** lepton numbers, $\mathbf{L_e}$ and $\mathbf{L_\mu}$.

All the leptons and lepton-neutrinos have their own **antiparticle** too — no surprises there. They have the **opposite charge** and **lepton numbers** to their matching particles. For example, the antimuon μ^+ has charge = +1, $L_e = 0$ and $L_\mu = -1$.

Name	Symbol	Charge	L_e	L_μ
electron	e^-	−1	+1	0
electron-neutrino	ν_e	0	+1	0
muon	μ^-	−1	0	+1
muon-neutrino	ν_μ	0	0	+1

Practice Questions

Q1 List the differences between a hadron and a lepton.

Q2 Which is the only stable baryon (probably)?

Q3 A particle collision at CERN produces 2 protons, 3 pions and 1 neutron. What is the total baryon number of these particles?

Q4 Which two particles have lepton number $L_\mu = +1$?

Exam Questions

Q1 List all the decay products of the neutron.
Explain why this decay cannot be due to the strong interaction. [3 marks]

Q2 Initially, the muon was incorrectly identified as a meson.
Explain why the muon is not a meson. [3 marks]

Q3 A sodium atom contains 11 electrons, 11 protons and 12 neutrons. Which row in the table contains the correct numbers of hadrons, baryons, mesons and leptons in a sodium atom? [1 mark]

	Hadrons	Baryons	Mesons	Leptons
A	12	11	12	11
B	23	23	0	11
C	23	0	23	23
D	12	11	0	23

Go back to the top of page 10 — do not pass GO, do not collect £200...

Do it. Go back and read it again. I promise — read these pages a few times and you'll start to see a pattern. There are hadrons that feel the strong force, leptons that don't. Hadrons are either baryons or mesons, and they're all weird except for those well-known baryons: protons and neutrons. There are loads of leptons, including good old electrons.

Quarks

Quarks may sound like a bizarre concept, but they weren't just made up willy-nilly. Large teams of scientists and engineers all over the world worked for years to come up with the info on these pages.

Quarks** are **Fundamental Particles

Quarks are the **building blocks** for **hadrons** (baryons and mesons). Antiparticles of hadrons are made from **antiquarks**.

1) To make **protons** and **neutrons** you only need two types of quark — the **up** quark (**u**) and the **down** quark (**d**).
2) An extra one called the **strange** quark (**s**) lets you make more particles with a property called **strangeness**.

***Strangeness** is Only **Conserved Some** of the Time*

1) **Strangeness**, like baryon number, is a **quantum number** (see p.10) — it can only take a certain set of values.
2) Strange particles, such as kaons, are **created** via the **strong** interaction but **decay** via the **weak** interaction.
3) Here's the catch — strangeness is **conserved** in the **strong interaction**, but **not** in the **weak interaction** (p.14).
4) That means strange particles are **always produced in pairs** (e.g. K^+ and K^-).
   One has a strangeness of +1, and the other has a strangeness of –1, so the overall strangeness of 0 is **conserved**.

Quarks** and **Antiquarks** have **Opposite Properties

The **antiquarks** have **opposite properties** to the quarks — as you'd expect.

QUARKS

Name	Symbol	Charge	Baryon number	Strangeness
up	u	$+{}^2/_3$	$+{}^1/_3$	0
down	d	$-{}^1/_3$	$+{}^1/_3$	0
strange	s	$-{}^1/_3$	$+{}^1/_3$	–1

ANTIQUARKS

Name	Symbol	Charge	Baryon number	Strangeness
anti-up	$\bar{u}$	$-{}^2/_3$	$-{}^1/_3$	0
anti-down	$\bar{d}$	$+{}^1/_3$	$-{}^1/_3$	0
anti-strange	$\bar{s}$	$+{}^1/_3$	$-{}^1/_3$	+1

Baryons** are Made from **Three Quarks

Evidence for quarks came from **hitting protons** with **high-energy electrons**.
The way the **electrons scattered** showed that there were **three concentrations of charge** (quarks) **inside** the proton.

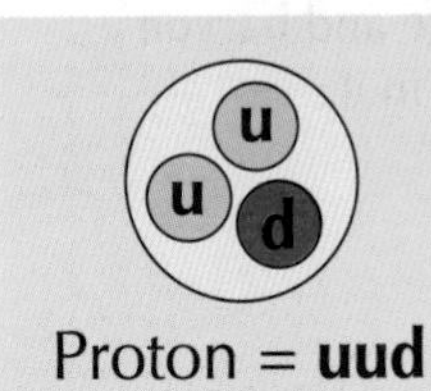

Proton = **uud**

Total charge
$= {}^2/_3 + {}^2/_3 - {}^1/_3 = 1$
Baryon number
$= {}^1/_3 + {}^1/_3 + {}^1/_3 = 1$

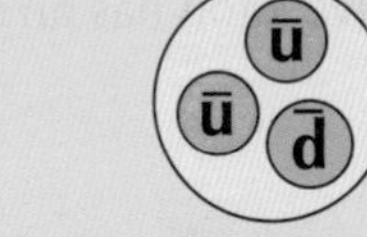

Antiproton = $\mathbf{\bar{u}\bar{u}\bar{d}}$

Total charge
$= -{}^2/_3 - {}^2/_3 + {}^1/_3 = -1$
Baryon number
$= -{}^1/_3 - {}^1/_3 - {}^1/_3 = -1$

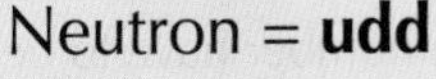

Neutron = **udd**

Total charge
$= {}^2/_3 - {}^1/_3 - {}^1/_3 = 0$
Baryon number
$= {}^1/_3 + {}^1/_3 + {}^1/_3 = 1$

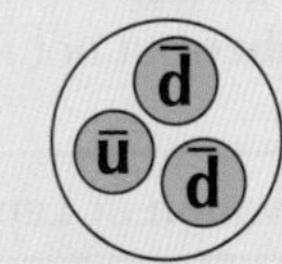

Antineutron = $\mathbf{\bar{u}\bar{d}\bar{d}}$

Total charge
$= -{}^2/_3 + {}^1/_3 + {}^1/_3 = 0$
Baryon number
$= -{}^1/_3 - {}^1/_3 - {}^1/_3 = -1$

Quarks

Mesons are a Quark and an Antiquark

Pions are just made from combinations of **up**, **down**, **anti-up** and **anti-down** quarks.
Kaons have **strangeness** so you need to put in **s** quarks as well (remember, the **s** quark has a strangeness of S = –1).

Before we move on, it's worth mentioning that the π^- meson is just the **antiparticle** of the π^+ meson, the **K⁻** meson is the antiparticle of the **K⁺** meson, and the **antiparticle** of a π^0 meson is **itself**. It all makes sense when you look at the quark compositions to the right...

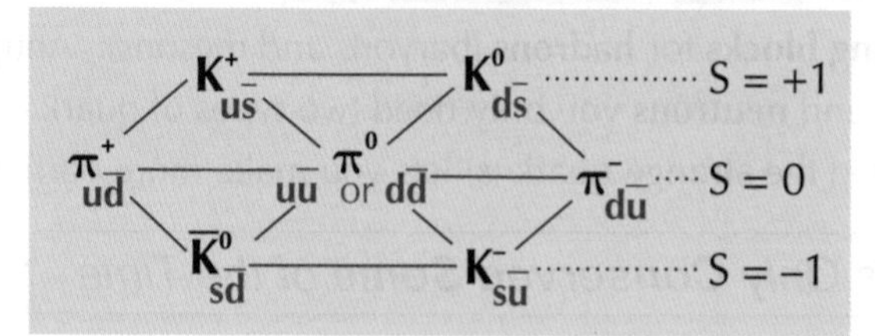

Physicists love patterns. Gaps in patterns like this predicted the existence of particles that were actually found later in experiments. Great stuff.

The Weak Interaction is something that Changes the Quark Type

In β^- decay a **neutron** is changed into a **proton** — in other words **udd** changes into **uud**. It means turning a **d** quark into a **u** quark. Only the weak interaction can do this.

Some unstable isotopes like **carbon-11** decay by β^+ emission. In this case a **proton** changes to a **neutron**, so a **u** quark changes to a **d** quark and we get:

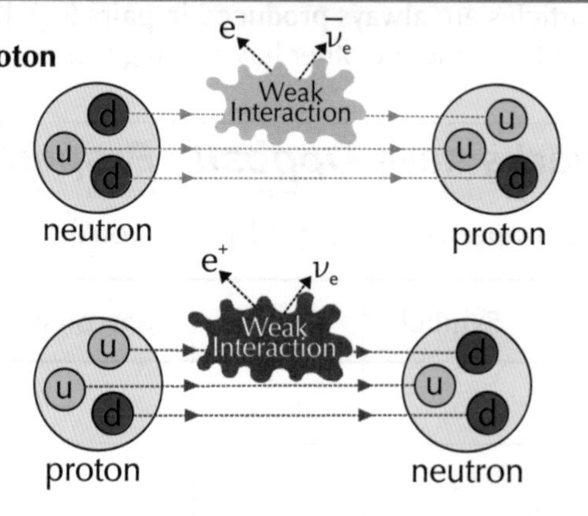

Four Properties are Conserved in Particle Interactions

Charge is Always Conserved

In **any** particle interaction, the **total charge** after the interaction must equal the total charge before the interaction.

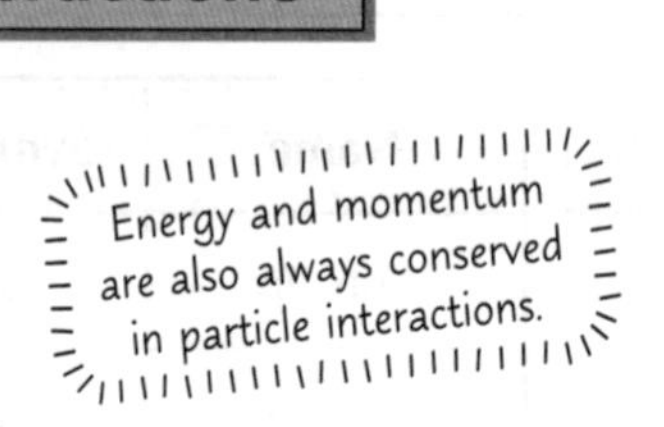

Baryon Number is Always Conserved

Just like with charge, in **any** particle interaction, the **baryon number** after the interaction must equal the baryon number before the interaction.

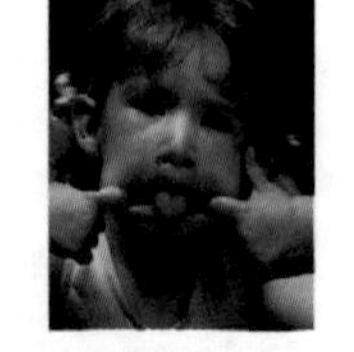
Dylan was committed to conserving strangeness.

Strangeness is Conserved in Strong Interactions

The **only** way to change the **type** of quark is with the **weak interaction**, so in strong interactions there has to be the same number of strange quarks at the beginning as at the end. In weak interactions, strangeness can change by –1, 0 or +1. The interaction $K^- + p \rightarrow n + \pi^0$ is fine for **charge** and **baryon number** but not for **strangeness** — so it won't happen. The negative kaon has an **s** quark in it.

Conservation of Lepton Number is a Bit More Complicated

The **different types** of lepton number have to be conserved **separately**.

1) For example, the interaction $\pi^- \rightarrow \mu^- + \bar{\nu}_\mu$ has $L_\mu = 0$ at the start and $L_\mu = 1 - 1 = 0$ at the end, so it's OK. Similarly, $n \rightarrow p + e^- + \bar{\nu}_e$ is fine. $L_e = 0$ at the start and $L_e = 1 - 1 = 0$ at the end.
2) On the other hand, the interaction $\nu_\mu + \mu^- \rightarrow e^- + \nu_e$ can't happen. At the start $L_\mu = 2$ and $L_e = 0$, but at the end $L_\mu = 0$ and $L_e = 2$.

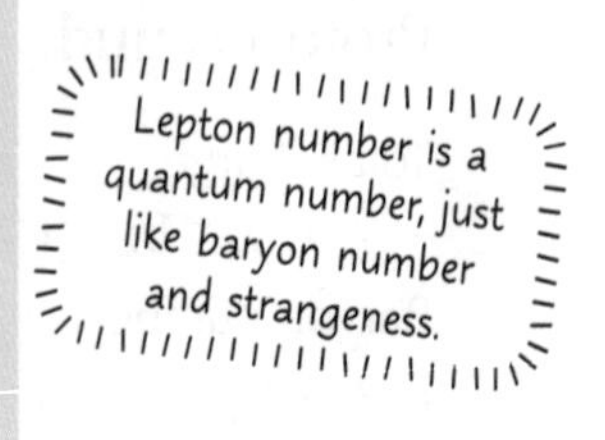

Quarks

There's No Such Thing as a Free Quark

What if you **blasted** a **proton** with **enough energy** — could you **separate out** the quarks? Nope. Your energy just gets changed into more **quarks and antiquarks** — it's **pair production** again and you just make **mesons**. It's not possible to get a quark by itself — this is called **quark confinement**.

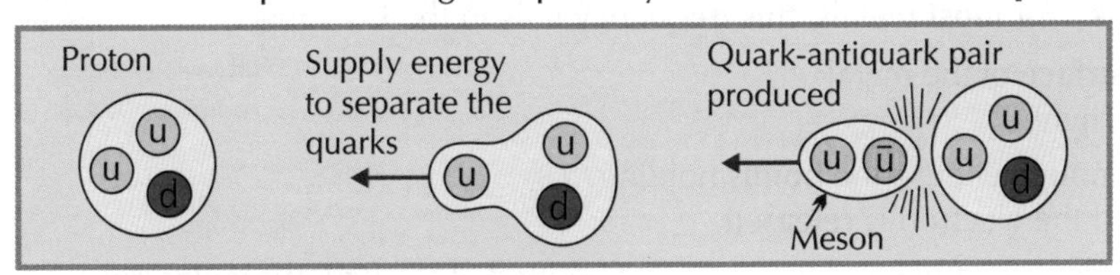

Simon was practising quack confinement.

We're Still Searching for Particles

As time goes on, our knowledge and understanding of particle physics **changes**.

1) **New theories** are created to try to explain observations from experiments. Sometimes, physicists hypothesise a new **particle** and the **properties** they expect it to have. E.g. the **neutrino** was hypothesised due to observations of beta decay.
2) **Experiments** to try to find the existence of this new particle are then carried out. Results from different experiments are **combined** to try to **confirm** the new particle. If it exists, the theory is **more likely** to be correct and the scientific community start to accept it — it's **validated**.
3) It's not quite that simple though. Experiments in particle physics often need particles travelling at incredibly **high speeds** (close to the speed of light). This can only be achieved using **particle accelerators**. These huge pieces of equipment are very **expensive** to build and run. This means that **large groups** of scientists and engineers from all over the **world** have to **collaborate** to be able to fund these experiments.

Example: Paul Dirac predicted the existence of **antimatter** in 1928. His theory was **validated** with the observation of the **positron** and, over the years, more and more observations of antiparticles. Nowadays, it's **accepted** that antimatter exists, but there are still **questions**. For example, there should have been **equal amounts** of matter and antimatter created when the universe was formed, but **almost everything** we observe is made of **matter**.

Scientists are trying to figure out what happened to all the antimatter by studying the differences in behaviour of matter and antimatter particles using the **Large Hadron Collider (LHC)** at **CERN**. The LHC is a **17 mile long** particle accelerator costing around **£3 billion** to build and **£15 million** per year to run. Some **10,000** scientists from **100** countries are involved.

ATLAS, just one of many experiments the LHC at CERN is used for, involves around 3000 scientists from 38 different countries.

Practice Questions

Q1 What is a quark?

Q2 Kaons are produced by the strong interaction. Why must they be produced in pairs?

Q3 By how much can the strangeness change in a weak interaction?

Q4 Which type of particle is made from a quark and an antiquark?

Q5 Describe how a neutron is made up from quarks.

Q6 List six quantities that are conserved in strong particle interactions.

Exam Questions

Q1 a) Write down the quark composition of the π^-. [1 mark]

b) Explain how the charges of the quarks give rise to its charge. [1 mark]

Q2 Explain how the quark composition is changed in the β^- decay of the neutron. [2 marks]

Q3 Give two reasons why the reaction $p + p \rightarrow p + K^+$ does not happen. [2 marks]

A physical property called strangeness — how cool is that...

True, there's a lot of information here, but this page really does tie up a lot of the stuff on the last few pages. Learn as much as you can from this three-page spread, then go back to page 13, and work back through to here. Don't expect to understand it all — but you'll definitely find it much easier to learn when you can see how all the bits fit together.

The Photoelectric Effect

I think they should rename 'the photoelectric effect' as 'the piece-of-cake effect' — it's not easy, I just like cake.

Shining Light on a Metal can Release Electrons

'Light' means any EM radiation — not just visible light.

If you shine **light** of a **high enough frequency** onto the **surface of a metal**, the metal will **emit electrons**. For **most** metals, this **frequency** falls in the **UV** range.

ultraviolet radiation

electrons

1) **Free electrons** on the **surface** of the metal **absorb energy** from the light.
2) If an electron **absorbs enough** energy, the **bonds** holding it to the metal **break** and the electron is **released**.
3) This is called the **photoelectric effect** and the electrons emitted are called **photoelectrons**.

You don't need to know the details of any experiments on this, you just need to learn the three main conclusions:

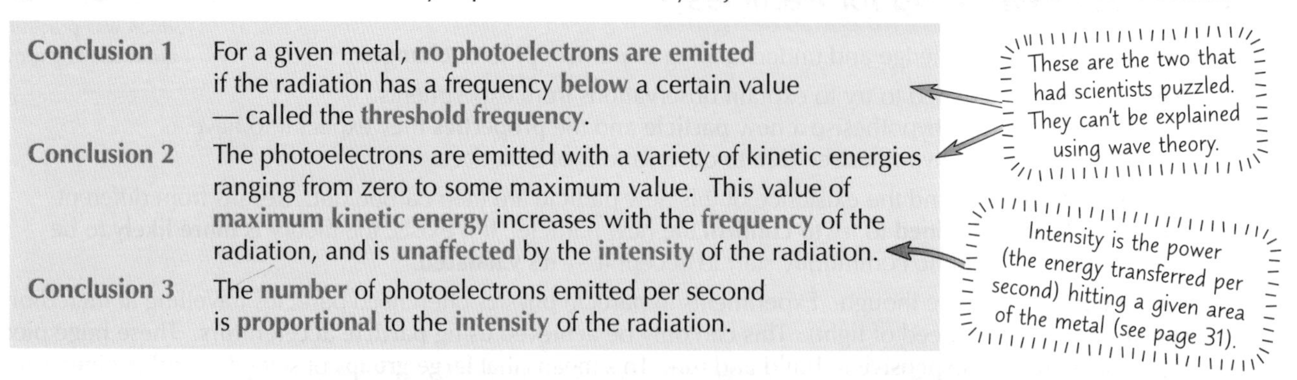

Conclusion 1 For a given metal, **no photoelectrons are emitted** if the radiation has a frequency **below** a certain value — called the **threshold frequency**.

Conclusion 2 The photoelectrons are emitted with a variety of kinetic energies ranging from zero to some maximum value. This value of **maximum kinetic energy** increases with the **frequency** of the radiation, and is **unaffected** by the **intensity** of the radiation.

Conclusion 3 The **number** of photoelectrons emitted per second is **proportional** to the **intensity** of the radiation.

The Photoelectric Effect Couldn't be Explained by Wave Theory...

According to wave theory:

1) For a particular frequency of light, the **energy** carried is **proportional** to the **intensity** of the beam.
2) The energy carried by the light would be **spread evenly** over the wavefront.
3) **Each** free electron on the surface of the metal would gain a **bit of energy** from each incoming wave.
4) Gradually, each electron would gain **enough energy** to leave the metal.

SO... The **higher the intensity** of the wave, the **more energy** it should transfer to each electron — the kinetic energy should increase with **intensity**. There's **no explanation** for the **kinetic energy** depending only on the **frequency**.

There is also **no explanation** for the **threshold frequency**. According to **wave theory**, the electrons should be emitted **eventually**, no matter what the **frequency** is.

...But it Could be Explained by Einstein's Photon Model of Light

1) **Einstein** suggested that **EM waves** (and the energy they carry) **exist** in discrete packets — called **photons**.
2) The **energy carried** by one of these **photons** is:

$$E = hf = \frac{hc}{\lambda}$$

where h = Planck's constant = 6.63×10^{-34} Js
and c = speed of light in a vacuum = 3.00×10^{8} ms^{-1}

You might have seen this formula before on page 6.

3) Einstein saw these photons of light as having a **one-on-one**, **particle-like** interaction with **an electron** in a **metal surface**. A photon would **transfer all** its **energy** to **one**, **specific electron**.

According to the photon model:

1) When light hits its surface, the metal is **bombarded** by photons.
2) If one of these photons **collides** with a free electron, the electron will gain energy equal to ***hf***.

Before an electron can **leave** the surface of the metal, it needs enough energy to **break the bonds holding it there**. This energy is called the **work function** (which has the symbol ϕ (phi)) and its **value** depends on the **metal**.

The Photoelectric Effect

The **Photon Model** Explains the **Threshold Frequency**...

1) If the energy **gained** by an electron (on the surface of the metal) from a photon is **greater** than the **work function**, the electron is **emitted**.
2) If it **isn't**, the metal will heat up, but **no electrons** will be emitted.
3) Since, for **electrons** to be released, $hf \geq \phi$, the **threshold frequency** must be: $f = \frac{\phi}{h}$

...and the **Maximum Kinetic Energy**

1) The **energy transferred** to an electron is hf.
2) The **kinetic energy** the electron will be carrying when it **leaves** the metal is hf **minus** any energy it's **lost** on the way out. Electrons **deeper** down in the metal lose more energy than the electrons on the **surface**, which explains the **range** of energies.
3) The **minimum** amount of energy it can lose is the **work function**, so the **maximum kinetic energy** of a photoelectron, $E_{k\,(max)}$, is given by the photoelectric equation:

$$hf = \phi + E_{k\,(max)} \quad \text{where} \quad E_{k\,(max)} = \frac{1}{2}mv_{max}^2$$

4) The **kinetic energy** of the electrons is **independent of the intensity** (the **number** of photons **per second** on an **area**, p.31), as they can **only absorb one photon** at a time. Increasing the **intensity** just means **more photons per second** on an **area** — each photon has the **same energy** as before.

The **Stopping Potential** Gives the **Maximum Kinetic Energy**

1) The **maximum kinetic energy** can be measured using the idea of **stopping potential**.
2) The **emitted electrons** are made to lose their energy by **doing work** against an applied **potential difference**.
3) The **stopping potential**, V_s, is the p.d. needed to stop the **fastest** moving electrons, with $E_{k\,(max)}$.
4) The **work done** by the p.d. in **stopping** the fastest electrons is equal to the **energy** they were carrying:

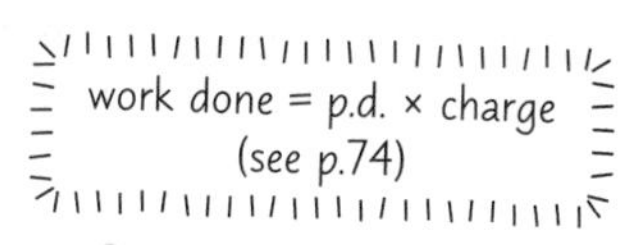

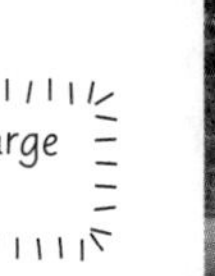

Work done = zero.

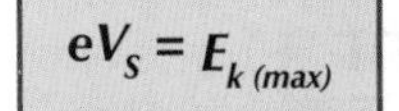

$$eV_s = E_{k\,(max)}$$

where e = charge on the electron = 1.60×10^{-19} C, V_s = stopping potential in V, and $E_{k\,(max)}$ is measured in J.

Practice Questions

Q1 Explain what the photoelectric effect is.

Q2 What three conclusions were drawn from experimentation on the photoelectric effect?

Q3 What is meant by the work function of a metal?

Q4 How is the maximum kinetic energy of a photoelectron related to the work function?

Q5 Explain what is meant by the stopping potential. Write down a formula relating stopping potential and $E_{k\,(max)}$.

Exam Questions

Q1 An isolated zinc plate with neutral charge is exposed to high-frequency ultraviolet light. State and explain the effect of the ultraviolet light on the charge of the plate. [2 marks]

Q2 Explain why photoelectric emission from a metal surface only occurs when the frequency of the incident radiation exceeds a certain threshold value. [2 marks]

I'm so glad we got that all cleared up...

The most important bits here are why wave theory doesn't explain the phenomenon of the photoelectric effect, and why the photon theory does. A good way to learn conceptual stuff like this is to try to explain it to someone else.

Energy Levels and Photon Emission

Quantum theory doesn't really make much sense — to anyone. It works though, so it's hard to argue with.

Electrons in Atoms Exist in Discrete Energy Levels

The ground state is the lowest energy state of the atom.

LEVEL	ENERGY
n = ∞	zero energy
n = 5	-8.6×10^{-20} J or -0.54 eV
n = 4	-1.4×10^{-19} J or -0.85 eV
n = 3	-2.4×10^{-19} J or -1.5 eV
n = 2	-5.4×10^{-19} J or -3.4 eV
n = 1	-2.2×10^{-18} J or -13.6 eV

transitions

1) **Electrons** in an **atom** can **only exist** in certain **well-defined energy levels.** Each level is given a **number**, with **n = 1** representing the **ground state.**
2) Electrons can **move down** energy levels by **emitting** a **photon.**
3) Since these **transitions** are between **definite energy levels**, the **energy** of **each photon** emitted can **only** take a **certain allowed value.**
4) The diagram on the right shows the **energy levels** for **atomic hydrogen.**
5) The **energies involved** are **so tiny** that it makes sense to use a more **appropriate unit** than the **joule.** The **electronvolt (eV)** is defined as:

> The **kinetic energy carried** by an **electron** after it has been **accelerated** through a **potential difference** of **1 volt.**

energy gained by electron (eV) = accelerating voltage (V)

$$1 \text{ eV} = 1.60 \times 10^{-19} \text{ J}$$

6) On the diagram, energies are labelled in **both units** for **comparison's** sake.
7) The **energy** carried by each **photon** is **equal** to the **difference in energies** between the **two levels.** The equation below shows a **transition** between a higher energy level n = 1 where the electrons have energy E_1 and a lower energy level n = 2 with electrons of energy E_2:

$$\Delta E = E_2 - E_1 = hf = \frac{hc}{\lambda}$$

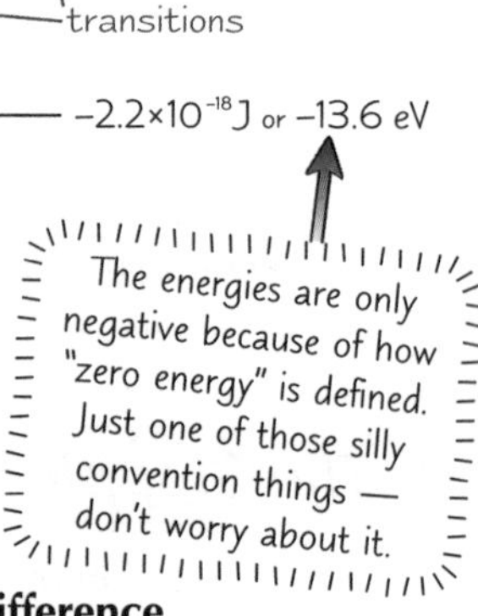

8) Electrons can also **move up** energy levels if they **absorb a photon** with the **exact energy difference** between the two levels. The movement of an electron to a higher energy level is called **excitation.**
9) If an electron is **removed** from an atom, we say the atom is **ionised.** The energy of each **energy level** within an atom gives the amount of **energy** needed to **remove an electron** in that level from the atom. The **ionisation energy** of an atom is the amount of energy needed to completely remove an electron from the atom from the **ground state** (**n = 1**).

Fluorescent Tubes use Excited Electrons to Produce Light

1) **Fluorescent tubes** contain **mercury vapour**, across which an initial **high voltage** is applied. This **high voltage** accelerates **fast-moving free electrons** that **ionise** some of the **mercury atoms**, producing **more** free electrons.
2) When this flow of free electrons **collides** with electrons in **other mercury atoms**, the electrons in the mercury atoms are **excited** to **higher energy levels.**
3) When these **excited electrons** return to their **ground states**, they emit **photons** in the **UV** range.
4) A **phosphorus coating** on the **inside** of the tube **absorbs** these **photons**, exciting its **electrons** to **much higher orbits.** These electrons then **cascade** down the **energy levels**, **emitting** many **lower energy photons** in the form of **visible light.**

Fluorescent Tubes Produce Line Emission Spectra

1) If you **split** the light from a **fluorescent tube** with a **prism** or a **diffraction grating** (see pages 34-35), you get a **line spectrum.**
2) A line spectrum is seen as a **series** of **bright lines** against a **black background.**
3) Each **line** corresponds to a **particular wavelength** of light **emitted** by the source.
4) Since only certain photon energies are allowed, you only see the **wavelengths** corresponding to these energies.

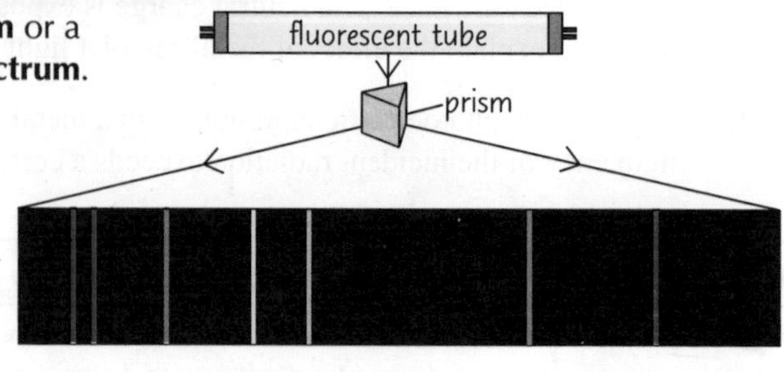

Energy Levels and Photon Emission

*Shining **White Light** through a **Cool Gas** gives an **Absorption Spectrum***

Continuous Spectra** Contain **All** Possible **Wavelengths

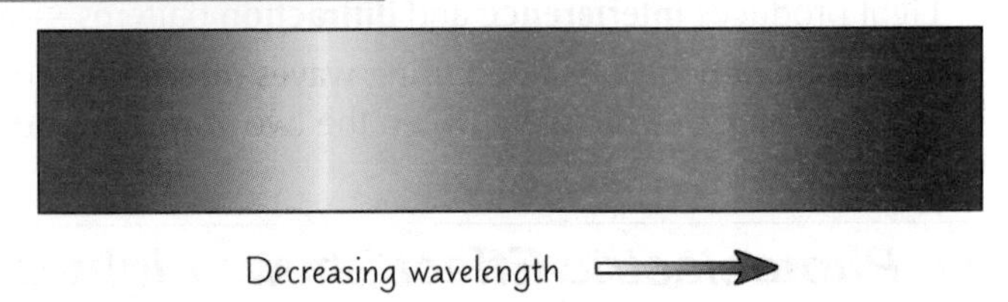

1) The **spectrum** of **white light** is **continuous**.
2) If you **split** the **light** up with a **prism**, the **colours** all **merge** into each other — there **aren't** any **gaps** in the spectrum.
3) **Hot things** emit a **continuous spectrum** in the visible and infrared.
4) **All** the **wavelengths** are allowed because the electrons are **not confined** to **energy levels** in the object producing the **continuous spectrum**. The electrons are not bound to atoms and are **free**.

***Cool** Gases **Remove** Certain **Wavelengths** from the Continuous Spectrum*

1) You get a **line absorption spectrum** when **light** with a **continuous spectrum** of **energy** (white light) passes through a cool gas.
2) At **low temperatures**, **most** of the **electrons** in the **gas atoms** will be in their **ground states**.
3) The electrons can only absorb **photons** with **energies** equal to the **difference** between **two energy levels**.
4) **Photons** of the **corresponding wavelengths** are **absorbed** by the **electrons** to **excite** them to **higher energy levels**.
5) These **wavelengths** are then **missing** from the **continuous spectrum** when it **comes out** the other side of the gas.
6) You see a **continuous spectrum** with **black lines** in it corresponding to the **absorbed wavelengths**.
7) If you **compare** the **absorption** and **emission spectra** of a **particular gas**, the **black lines** in the **absorption spectrum match up** to the **bright lines** in the **emission spectrum**.

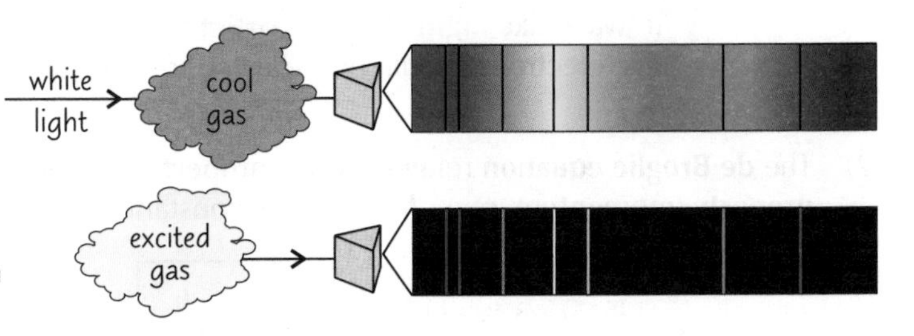

Practice Questions

Q1 Describe line absorption and line emission spectra. How are these two types of spectra produced?

Q2 Use the size of the energy level transitions involved to explain how the coating on a fluorescent tube converts UV into visible light.

Exam Question

Q1 An electron is accelerated through a potential difference of 12.1 V.

a) How much kinetic energy has it gained in i) eV and ii) joules? [2 marks]

b) This electron hits a hydrogen atom in its ground state and excites it.

i) Explain what is meant by excitation. [1 mark]

ii) Using the energy values on the right, calculate which energy level the electron from the hydrogen atom is excited to. [1 mark]

iii) Calculate the energies of the three photons that might be emitted as the electron returns to its ground state. [3 marks]

n = 5 ——— – 0.54 eV
n = 4 ——— – 0.85 eV
n = 3 ——— – 1.5 eV
n = 2 ——— – 3.4 eV
n = 1 ——— – 13.6 eV

I can honestly say I've never got so excited that I've produced light...

This is heavy stuff. Quite interesting though, as I was just saying to Dom a moment ago. He's doing a psychology book. Psychology's probably quite interesting too — and easier. But it won't help you become an astrophysicist.

Wave-Particle Duality

Is it a wave? Is it a particle? No, it's a wave. No, it's a particle. No it's not, it's a wave. No, don't be daft, it's a particle.

Interference and *Diffraction* show *Light* as a *Wave*

1) Light produces **interference** and **diffraction** patterns — **alternating bands** of **dark** and **light**.
2) These can **only** be explained using **waves interfering constructively** (when two waves overlap in phase) or **interfering destructively** (when the two waves are out of phase). (See p.26.)

The *Photoelectric Effect* Shows *Light* Behaving as a *Particle*

1) **Einstein** explained the results of **photoelectricity experiments** (see p.16) by thinking of the **beam of light** as a series of **particle-like photons**.
2) If a **photon** of light is a **discrete** bundle of energy, then it can **interact** with an **electron** in a **one-to-one way**.
3) **All** the **energy** in the **photon** is **given** to one **electron**.

De Broglie Came Up with the *Wave-Particle Duality Theory*

1) Louis de Broglie made a **bold suggestion** in his **PhD thesis:**

> If **'wave-like' light** showed **particle properties** (photons), **'particles'** like **electrons** should be expected to show **wave-like properties**.

I'm not impressed — this is just speculation. What do you think, Dad?

2) The **de Broglie equation** relates a **wave property** (**wavelength**, λ) to a **moving particle property** (**momentum**, mv). h = Planck's constant = 6.63×10^{-34} Js.

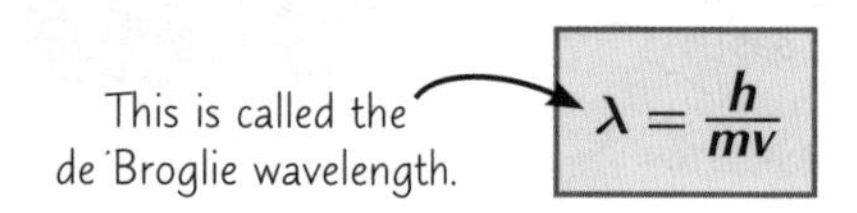

This is called the de Broglie wavelength.

$$\lambda = \frac{h}{mv}$$

3) The **de Broglie wave** of a particle can be interpreted as a **'probability wave'**. (The probability of finding a particle at a point is directly proportional to the square of the amplitude of the wave at that point — but you don't need to know that for your exam.)
4) Many physicists at the time **weren't very impressed** — his ideas were just **speculation**. But later experiments **confirmed** the wave nature of electrons.

Electron Diffraction shows the *Wave Nature* of *Electrons*

1) **Diffraction patterns** are observed when **accelerated electrons** in a vacuum tube **interact** with the **spaces** in a graphite **crystal**.
2) This **confirms** that electrons show **wave-like** properties.
3) According to wave theory, the **spread** of the **lines** in the diffraction pattern **increases** if the **wavelength** of the wave is **greater**.
4) In electron diffraction experiments, a **smaller accelerating voltage**, i.e. **slower** electrons, gives more **widely-spaced** rings.
5) **Increase** the **electron speed** (and therefore the electron **momentum**) and the diffraction pattern circles **squash together** towards the **middle**. This fits in with the **de Broglie** equation above — if the **momentum** is **greater**, the **wavelength** is **shorter** and the **spread** of the lines is **smaller**.

> In general, λ for **electrons** accelerated in a **vacuum tube** is about the **same size** as **electromagnetic waves** in the **X-ray** part of the spectrum.

6) If particles with a **greater mass** (e.g. **neutrons**) were travelling at the **same speed** as the electrons, they would show a more **tightly-packed diffraction pattern**. That's because a neutron's **mass** (and therefore its **momentum**) is **much greater** than an electron's, and so a neutron has a **shorter de Broglie wavelength**.

Wave-Particle Duality

***Particles Don't** show **Wave-Like Properties** All the Time*

You **only** get **diffraction** if a particle interacts with an object of about the **same size** as its **de Broglie wavelength**. A **tennis ball**, for example, with **mass 0.058 kg** and **speed 100 ms^{-1}** has a **de Broglie wavelength** of **10^{-34} m**. That's **10^{19} times smaller** than the **nucleus** of an **atom**! There's nothing that small for it to interact with.

Example: An electron of mass 9.11×10^{-31} kg is fired from an electron gun at 7.00×10^{6} ms^{-1} (to 3 s.f.). What size object will the electron need to interact with in order to diffract?

Momentum of electron = $mv = (9.11 \times 10^{-31}) \times (7.00 \times 10^{6}) = 6.377 \times 10^{-24}$ kg ms^{-1}

$\lambda = h/mv = 6.63 \times 10^{-34} / 6.377 \times 10^{-24} = 1.0396... \times 10^{-10} =$ **1.04×10^{-10} m (to 3 s.f.)**

Only crystals with atom layer spacing around this size are likely to cause the diffraction of this electron.

***Wave-Particle Duality** Wasn't Accepted Straight Away*

De Broglie first **hypothesised** wave-particle duality to explain **observations** of light acting as both a particle and a wave. But his theory **wasn't accepted** straight away. **Other scientists** had to **evaluate** de Broglie's theory (by a process known as **peer review**) before he **published** it, and then it was **tested with experiments**. Once enough evidence was found to back it up, the theory was accepted as **validated** by the scientific community.

Scientists' understanding of the nature of matter has changed over time through this process of **hypothesis and validation**. De Broglie's theory is **accepted** to be true — that is, until any new conflicting evidence comes along.

Practice Questions

Q1 Which observations show light to have a 'wave-like' character?

Q2 Which observations show light to have a 'particle' character?

Q3 What happens to the de Broglie wavelength of a particle if its momentum increases?
How does this affect the particle's diffraction pattern?

Q4 Particle A has a de Broglie wavelength of 8×10^{-10} m and particle B has a de Broglie wavelength of 2×10^{-10} m.
If the particles are travelling at the same speed, which particle has the greater mass?

Q5 Which observations show electrons to have a 'wave-like' character?

Exam Questions

Q1 a) State what is meant by the wave-particle nature of electromagnetic radiation. [1 mark]

b) Calculate the momentum of an electron with a de Broglie wavelength of 590 nm. [2 marks]

Q2 Electrons travelling at a speed of 3.50×10^{6} ms^{-1} exhibit wave properties.

a) Calculate the wavelength of these electrons. (Mass of an electron = 9.11×10^{-31} kg) [2 marks]

b) Calculate the speed of protons with the same wavelength as these electrons.
(Mass of a proton = 1.67×10^{-27} kg) [2 marks]

c) Some electrons and protons were accelerated from rest by the same potential difference, giving them the same kinetic energy. Explain why they will have different wavelengths. [3 marks]

Don't hide your wave-particles under a bushel...

Right — I think we'll all agree that quantum physics is a wee bit strange when you come to think about it. What it's saying is that electrons and photons aren't really waves, and they aren't really particles — they're both... at the same time. It's what quantum physicists like to call a 'juxtaposition of states'. Well they would, wouldn't they...

Progressive Waves

Aaaah... playing with long springs and waggling ropes about. It's all good clean fun as my mate Richard used to say...

A ***Wave*** *is the* ***Oscillation*** *of* ***Particles*** *or* ***Fields***

A **progressive** (moving) wave carries **energy** from one place to another **without transferring any material**.
A wave is caused by something making particles or fields oscillate (or vibrate) at a source.
These oscillations pass through the medium (or field) as the wave travels, carrying energy with it.

Here are some ways you can tell waves carry energy:

1) Electromagnetic waves cause things to **heat up**.
2) **X-rays** and **gamma rays** knock electrons out of their orbits, causing **ionisation**.
3) Loud **sounds** cause large oscillations of air particles which can make things **vibrate**.
4) **Wave power** can be used to **generate electricity**.
5) Since waves carry energy away, the **source** of the wave **loses energy**.

Alex loved wave power.

You Need to Know These ***Bits*** *of a* ***Wave***

1) **Cycle** — one **complete vibration** of the wave.
2) **Displacement**, ***x***, metres — how far a **point** on the wave has **moved** from its **undisturbed position.**
3) **Amplitude**, ***A***, metres — **maximum magnitude** of **displacement**.
4) **Wavelength**, **λ**, metres — the **length** of **one whole wave cycle**, from **crest** to **crest** or **trough** to **trough**.

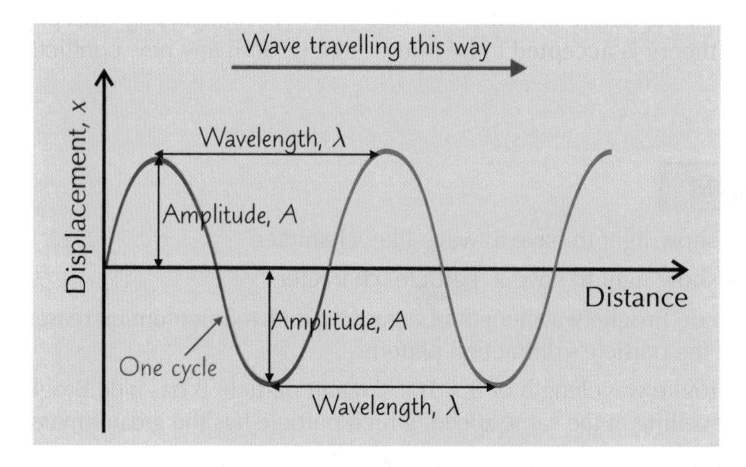

5) **Period**, ***T***, seconds — the **time taken** for a **whole cycle** (vibration) to complete, or to pass a given point.
6) **Frequency**, ***f***, hertz — the **number** of **cycles** (vibrations) **per second** passing a given **point**.
7) **Phase** — a measurement of the **position** of a certain **point** along the wave cycle.
8) **Phase difference** — the amount one wave lags behind another.

Phase and phase difference are measured in **angles** (in degrees or radians) or as **fractions of a cycle** (see p.26).

Waves Can Be ***Reflected*** *and* ***Refracted***

Reflection — the wave is **bounced back** when it **hits a boundary**. E.g. you can see the reflection of light in mirrors. The reflection of water waves can be demonstrated in a ripple tank.

Refraction — the wave **changes direction** as it enters a **different medium**.
The change in direction is a result of the wave slowing down or speeding up.

The ***Frequency*** *is the* ***Inverse*** *of the* ***Period***

$$\textbf{Frequency} = \frac{\textbf{1}}{\textbf{Period}}$$

$$f = \frac{1}{T}$$

It's that simple.
Get the **units** straight: **1 Hz = 1 s^{-1}**.

Progressive Waves

The Wave Equation Links Wave Speed, Frequency and Wavelength

Wave speed can be measured just like the speed of anything else:

$$\text{Wave speed } (c) = \frac{\text{Distance travelled } (d)}{\text{Time taken } (t)}$$

Remember, you're not measuring how fast a physical point (like one molecule of rope) moves. You're measuring how fast a point on the **wave pattern** moves.

From this you can get to the **wave equation**, which you've seen before on p.6.

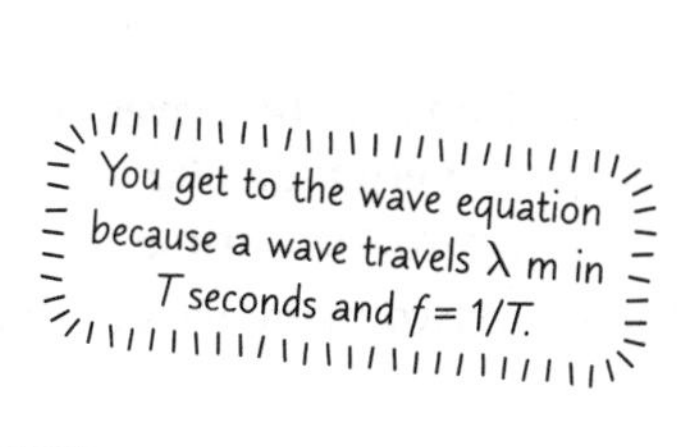

Speed of wave (c) = wavelength (λ) × frequency (f) $c = \lambda f$

Example: A wave has a wavelength of 420 m and travels at a speed of 125 ms^{-1}. Find the frequency of this wave.

Just rearrange the wave equation, to find f.

$$c = \lambda f \quad \text{so} \quad f = \frac{c}{\lambda} = \frac{125}{420} = 0.2976... = \mathbf{0.30\ Hz\ (to\ 2\ s.f.)}$$

The lowest no. of significant figures the question data is given to is 2 s.f., so give your answer to the same amount of significant figures.

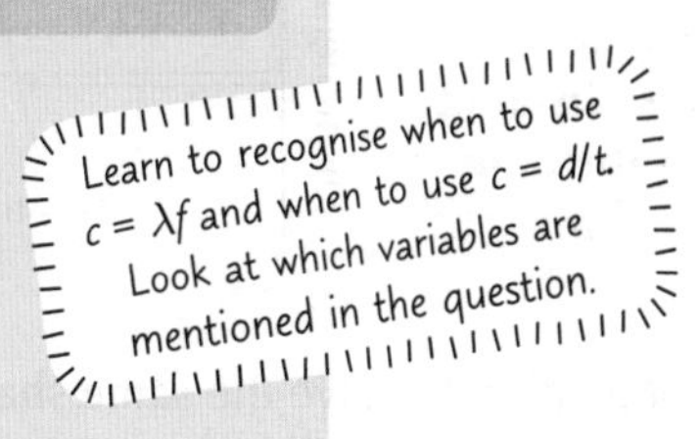

You may have seen **c** used before as the **speed of light** in a vacuum. Light is a type of electromagnetic wave. All EM waves travel with a **constant speed** in a **vacuum** of $c = 3.00 \times 10^8$ ms^{-1}. The c used in the wave equation is the speed of the **wave in question** — it can take **any** value depending on the wave.

Practice Questions

Q1 Does a wave carry matter **or** energy from one place to another?

Q2 Write down the relationship between the frequency of a wave and its time period.

Q3 Give the units of frequency, displacement and amplitude.

Q4 Write down the equation connecting c, λ and f.

Exam Questions

Q1 A buoy floating on the sea takes 6.0 seconds to rise and fall once (complete a full cycle). The difference in height between the buoy at its lowest and highest points is 1.2 m, and waves pass it at a speed of 3.0 ms^{-1}.

a) Calculate the wavelength. [2 marks]

b) State the amplitude of the waves. [1 mark]

Q2 Light travelling through a vacuum has a wavelength of 7.1×10^{-7} m. Calculate its frequency. [1 mark]

Q3 Which of the following statements is correct? [1 mark]

A	Progressive waves transfer energy by transferring material.
B	Progressive waves transfer energy by oscillating particles/fields.
C	The source of a progressive wave has a constant energy.
D	Light is faster than other EM waves in a vacuum.

Hope you haven't phased out...

This isn't too difficult to start you off — most of it you'll have done at GCSE anyway. But once again, it's a whole lot of definitions and a handy equation to remember, and you won't get far without learning them. Yada yada.

Longitudinal and Transverse Waves

There are different types of wave — the difference is easiest to see using a long spring. Try it — you'll have hours of fun.

Transverse Waves Vibrate at **Right Angles** to the Direction of **Energy Transfer**

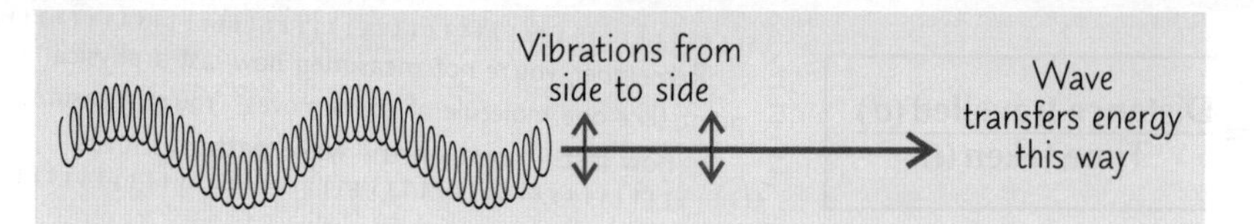

All **electromagnetic waves** are **transverse**. They travel as vibrations through magnetic and electric **fields** — with vibrations **perpendicular** to the direction of **energy transfer**. Other examples of transverse waves are **ripples** on water or waves on **strings**.

There are **two** main ways of **drawing** transverse waves:

1) They can be shown as **graphs** of **displacement** against **distance along the path of the wave.**

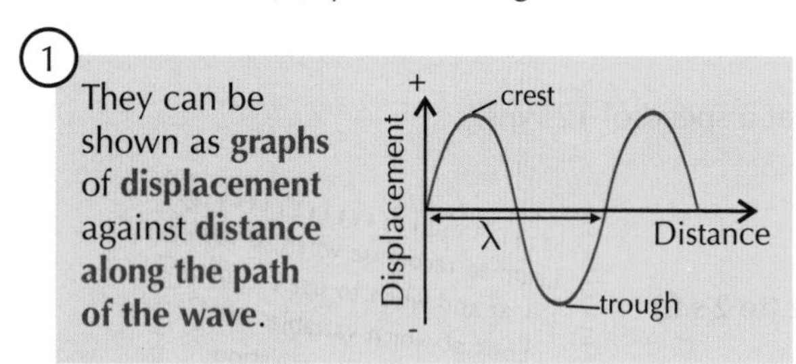

2) Or, they can be shown as graphs of **displacement against time** for a point as the wave passes.

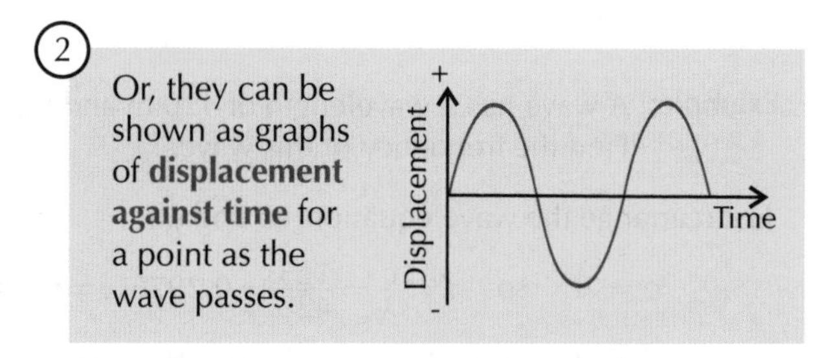

Both sorts of graph often give the **same shape**, so make sure you check out the label on the **x-axis**. Displacements **upwards** from the centre line are given a **+ sign**. Displacements downwards are given a **– sign**.

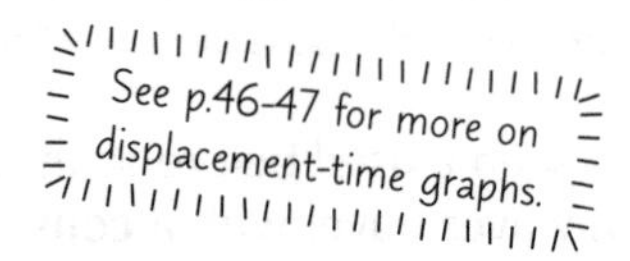
See p.46-47 for more on displacement-time graphs.

Longitudinal Waves Vibrate **Along** the Direction of **Energy Transfer**

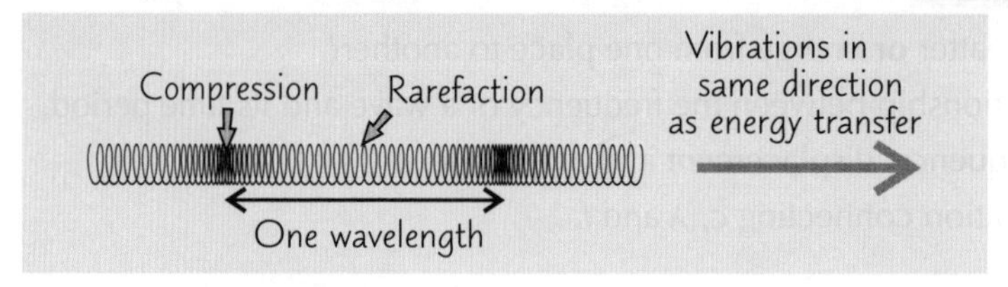

The most common example of a **longitudinal wave** is **sound**. A sound wave consists of alternate **compressions** and **rarefactions** of the **medium** it's travelling through. (That's why sound can't go through a vacuum.) Some types of earthquake shock waves are also longitudinal.

It's hard to **represent** longitudinal waves **graphically**. You'll usually see them plotted as **displacement** against **time**. These can be **confusing** though, because they look like a **transverse wave**.

A **Polarised Wave** Only **Oscillates** in One Direction

1) If you **shake a rope** to make a **wave** you can move your hand **up and down** or **side to side** or in a **mixture** of directions — it still makes a **transverse wave.**
2) But if you try to pass **waves in a rope** through a **vertical fence**, the wave will only get through if the **vibrations** are **vertical**. The fence filters out vibration in other directions. This is called **polarising** the wave.

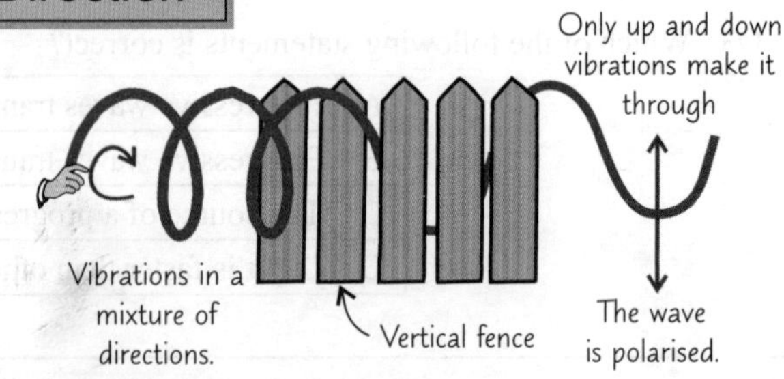

Polarisation **can only happen** for **transverse** waves.

Longitudinal and Transverse Waves

Polarisation is Evidence that Electromagnetic Waves are Transverse

In 1808, Etienne-Louis Malus discovered that light was **polarised** by reflection. Physicists at the time thought that light spread like sound, as a **longitudinal** wave, so they struggled to explain polarisation. In 1817, Young suggested light was a **transverse** wave consisting of vibrating electric and magnetic **fields** at right angles to the transfer of energy. This explained why light could be **polarised**.

Polarising Filters Only Transmit Vibrations in One Direction

1) Ordinary **light waves** are a mixture of **different directions** of **vibration**. (The things vibrating are electric and magnetic fields.) They can be **polarised** using a **polarising filter**.
2) If you have two polarising filters at **right angles** to each other, then **no** light will get through.
3) Light becomes **partially** polarised when reflected from some surfaces — some of it vibrates in the **same direction**.
4) If you view reflected partially polarised light through a polarising filter at the correct angle you can block out unwanted **glare**. **Polaroid sunglasses** make use of this effect.

Unpolarised light
Polarising filters
Transmission axes (light vibrating this way gets through)
Direction of vibrations

Television and Radio Signals are Polarised

If you walk down the street and look up at the **TV aerials** on people's houses, you'll see that the **rods** (the sticky-out bits) on them are all **horizontal**. The reason for this is that **TV signals** are **polarised** by the orientation of the **rods** on the **broadcasting aerial**. To receive a strong signal, you have to **line up** the **rods** on the **receiving aerial** with the **rods** on the **transmitting aerial** — if they aren't aligned, the signal strength will be lower.

It's the **same** with **radio** — if you try **tuning a radio** and then **moving** the **aerial** around, your signal will **come and go** as the transmitting and receiving aerials go in and out of **alignment**.

The rods on this broadcasting aerial are horizontal.

Practice Questions

Q1 Give an example of a transverse wave and a longitudinal wave.

Q2 What is a polarised wave?

Q3 How can you polarise light?

Q4 Why do you have to line up transmitting and receiving television aerials?

Exam Questions

Q1 Sunlight reflected from road surfaces mostly vibrates in one direction.

a) Explain how this is evidence that sunlight is made up of transverse waves. [2 marks]

b) Explain how Polaroid sunglasses help to reduce glare caused by reflections. [2 marks]

Q2 Explain why sound waves cannot be polarised. [2 marks]

So many waves — my arms are getting tired...

Right, there's lots to learn on these two pages, so I won't hold you up with chat. Don't panic though — a lot of this stuff will be familiar from GCSE, so it's not like you're starting from scratch. One last thing — I know television is on this page, but it doesn't mean you can tune in and call it revision — it won't help. Get the revision done, then take a break.

Superposition and Coherence

When two waves get together, it can be either really impressive or really disappointing.

Superposition *Happens When* ***Two*** *or* ***More*** *Waves* ***Pass Through*** *Each Other*

1) At the **instant** the waves **cross**, the **displacements** due to each wave **combine**. Then **each wave** goes on its merry way. You can **see** this if **two pulses** are sent **simultaneously** from each end of a rope.
2) The **principle of superposition** says that when two or more **waves cross**, the **resultant** displacement equals the **vector sum** of the **individual** displacements.

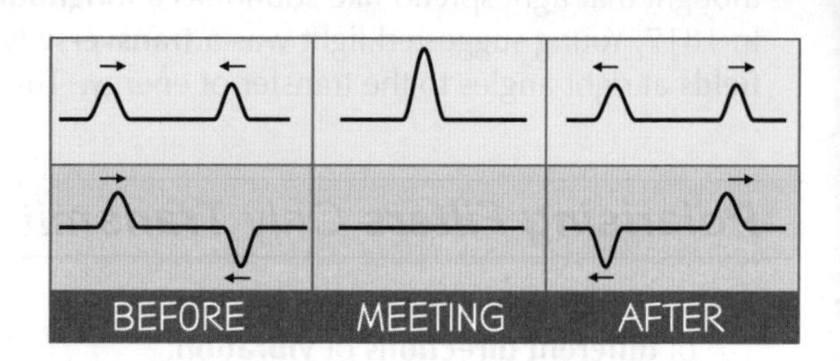

'Superposition' means 'one thing on top of another thing'. You can use the same idea in reverse — a complex wave can be separated out mathematically into several simple sine waves of various sizes.

Interference *can be* ***Constructive*** *or* ***Destructive***

1) A **crest** plus a **crest** gives a **bigger crest**. A **trough** plus a **trough** gives a **bigger trough**. These are both examples of **constructive interference**.
2) A **crest** plus a **trough** of **equal size** gives... **nothing**. The two displacements **cancel each other out** completely. This is called **destructive interference**.
3) If the **crest** and the **trough** aren't the **same size**, then the destructive interference **isn't total**. For the interference to be **noticeable**, the two **amplitudes** should be **nearly equal**.

Graphically, you can superimpose waves by adding the individual displacements at each point along the x-axis, and then plotting them.

In ***Phase*** *Means In* ***Step*** *— Two Points* ***In Phase*** *Interfere* ***Constructively***

1) Two points on a wave are **in phase** if they are both at the **same point** in the **wave cycle**. Points in phase have the **same displacement** and **velocity**.
   On the graph below, points **A** and **B** are **in phase**; points **A** and **C** are **out of phase**.
2) It's mathematically **handy** to show one **complete cycle** of a wave as an **angle of 360° (2π radians)**.

To convert from degrees to radians, multiply by π/180°.
To convert from radians to degrees, multiply by 180°/π.

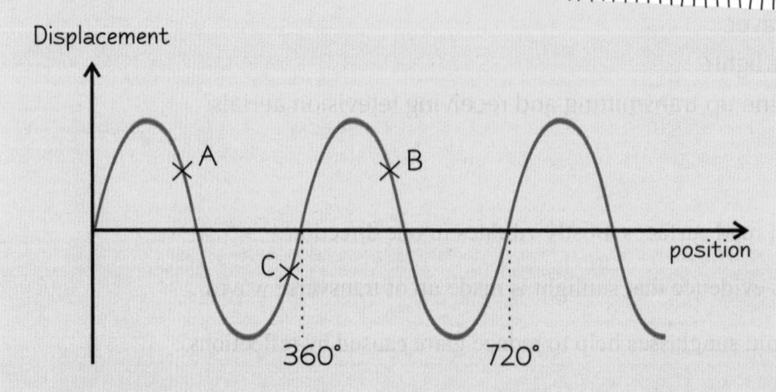

3) **Two points** with a **phase difference** of **zero** or a **multiple of 360°** (i.e. a **full cycle**) are **in phase**.
4) **Points** with a **phase difference** of **odd-number multiples** of **180°** (**π radians**, or a **half cycle**) are **exactly out of phase**.
5) You can also talk about two **different waves** being **in phase**. **In practice** this happens because **both** waves came from the **same oscillator**. In **other** situations there will nearly always be a **phase difference** between two waves.

Superposition and Coherence

To Get Interference Patterns the Two Sources Must Be Coherent

Interference **still happens** when you're observing waves of **different wavelength** and **frequency** — but it happens in a **jumble**. In order to get clear **interference patterns**, the two or more sources must be **coherent**.

Two sources are **coherent** if they have the **same wavelength** and **frequency** and a **fixed phase difference** between them.

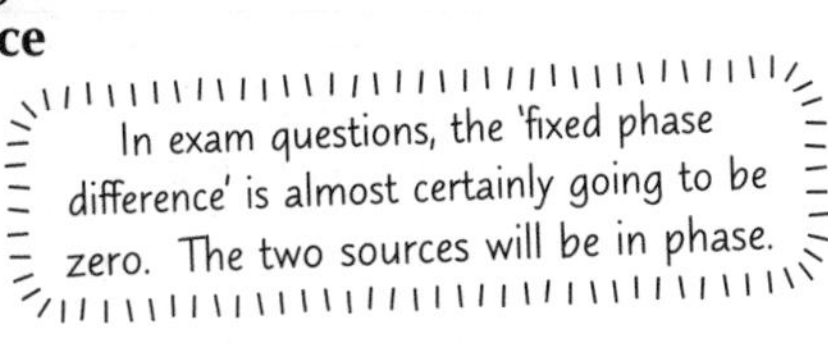

Constructive or Destructive Interference Depends on the Path Difference

1) Whether you get **constructive** or **destructive** interference at a **point** depends on how **much further one wave** has travelled than the **other wave** to get to that point.
2) The **amount** by which the path travelled by one wave is **longer** than the path travelled by the other wave is called the **path difference**.
3) At **any point an equal distance** from both sources you will get **constructive interference**. You also get constructive interference at any point where the **path difference** is a **whole number of wavelengths**. At these points the two waves are **in phase** and **reinforce** each other. But at points where the path difference is **half a wavelength**, **one and a half** wavelengths, **two and a half** wavelengths etc., the waves arrive **out of phase** and you get **destructive interference**.

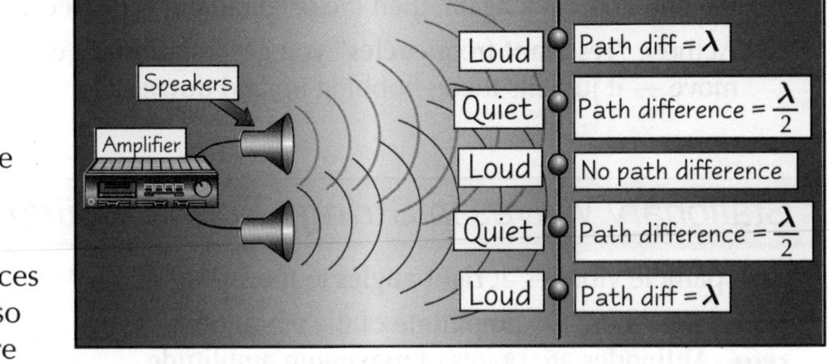

Constructive interference occurs when: **path difference = $n\lambda$ (where n is an integer)**

Destructive interference occurs when: **path difference = $\frac{(2n + 1)\lambda}{2} = (n + \frac{1}{2})\lambda$**

Practice Questions

Q1 Why does the principle of superposition deal with the **vector** sum of two displacements?

Q2 What happens when a crest meets a slightly smaller trough?

Q3 What is meant by the path difference of two waves?

Q4 If two points on a wave have a phase difference of 1440°, are they in phase?

Exam Questions

Q1 a) Two wave sources are coherent. Explain what this means. [2 marks]

b) Explain why you might have difficulty in observing interference patterns in an area affected by two waves from two sources even though the two sources are coherent. [1 mark]

Q2 Two points on a wave are exactly out of phase. Which row of the table correctly compares the two points? [1 mark]

	Phase Difference	Velocities	Displacements
A	180°	Equal	Opposite
B	180°	Opposite	Opposite
C	360°	Equal	Equal
D	360°	Opposite	Opposite

Learn this and you'll be in a super position to pass your exam...

...I'll get my coat.

There are a few really crucial concepts here: a) interference can be constructive or destructive, b) constructive interference happens when the path difference is a whole number of wavelengths, c) the sources must be coherent.

Stationary Waves

Stationary waves are weird things — they move but they don't actually go anywhere.

Progressive Waves Reflected at a Boundary Can Create a Stationary Wave

A stationary (standing) wave is the **superposition** of **two progressive waves** with the **same frequency** (**wavelength**), moving in **opposite directions**.

1) Unlike progressive waves, **no energy** is transmitted by a stationary wave.
2) You can demonstrate stationary waves by setting up a **driving oscillator** at one end of a **stretched string** with the other end fixed. The wave generated by the oscillator is **reflected** back and forth.
3) For most frequencies the resultant **pattern** is a **jumble**. However, if the oscillator happens to produce an **exact number of waves** in the time it takes for a wave to get to the **end** and **back again**, then the **original** and **reflected** waves **reinforce** each other.
4) At these **"resonant frequencies"** you get a **stationary wave** where the **pattern doesn't move** — it just sits there, bobbing up and down. Happy, at peace with the world...

A sitting wave.

Stationary Waves in Strings Form Oscillating "Loops" Separated by Nodes

Each particle vibrates at **right angles** to the string.

Nodes are where the **amplitude** of the vibration is **zero**. **Antinodes** are points of **maximum amplitude**.

At resonant frequencies, an **exact number** of **half wavelengths** fits onto the string:

First Harmonic

This stationary wave is vibrating at the **lowest possible** resonant frequency. It has **one "loop"** with a **node at each end**.

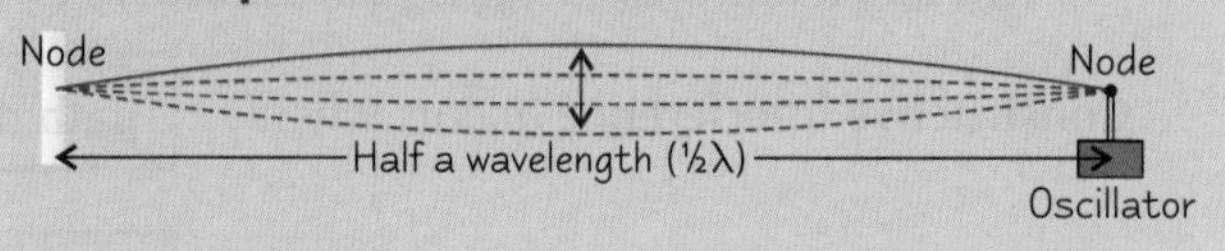

Second Harmonic

This is the **second harmonic**. It is **twice** the frequency of the **first harmonic**. There are two **"loops"** with a **node** in the **middle** and **one at each end**.

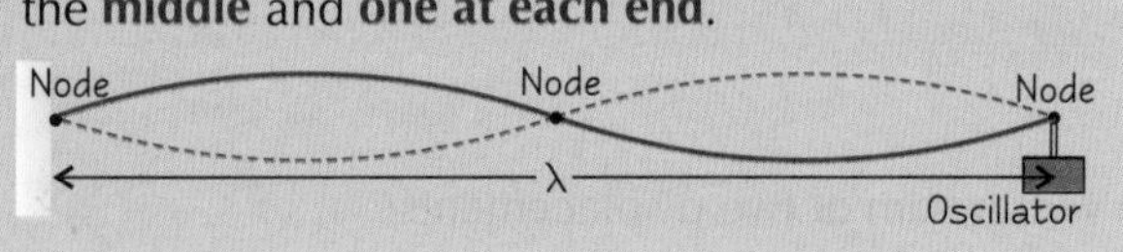

Third Harmonic

The **third harmonic** is **three times** the frequency of the first harmonic. **1½ wavelengths** fit on the string.

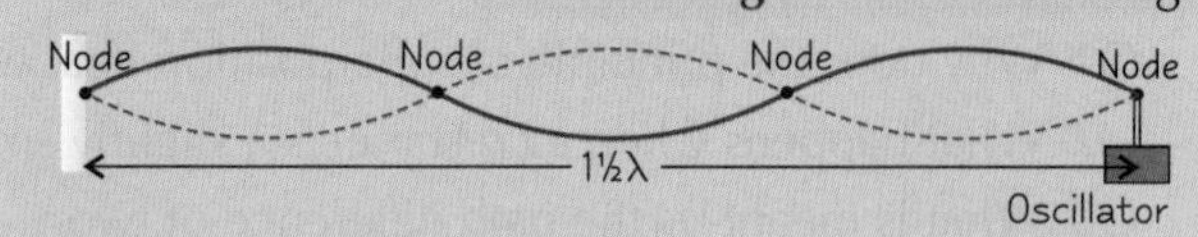

You can Demonstrate Stationary Waves with Microwaves and Sounds

Microwaves Reflected Off a Metal Plate Set Up a Stationary Wave

Microwave stationary wave apparatus

You can find the **nodes** and **antinodes** by moving the **probe** between the **transmitter** and **reflecting plate**.

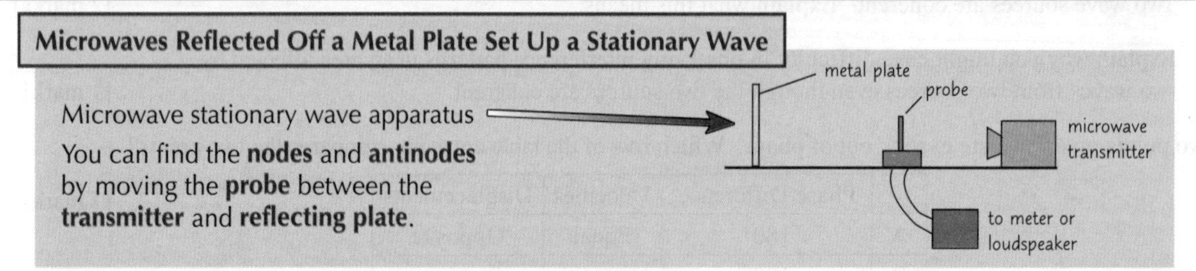

Powder Can Show Stationary Waves in a Tube of Air

Stationary **sound** waves are produced in the **glass tube**.

The lycopodium **powder** (don't worry, you don't need to know what that is) laid along the bottom of the tube is **shaken away** from the **antinodes** but left **undisturbed** at the **nodes**.

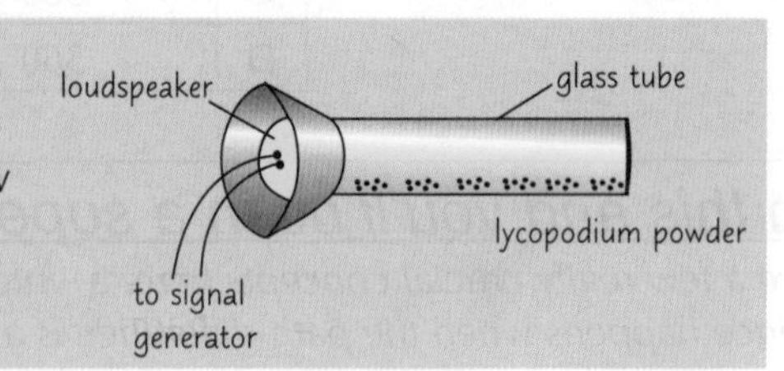

Stationary Waves

You Can Investigate **Factors** Affecting the **Resonant Frequencies** of a String

1) Start by measuring the **mass** (M) and **length** (L) of strings of different types using a **mass balance** and a ruler. Then find the **mass per unit length** of each string (μ) using: $\mu = \frac{M}{L}$ The units of μ are kgm^{-1}

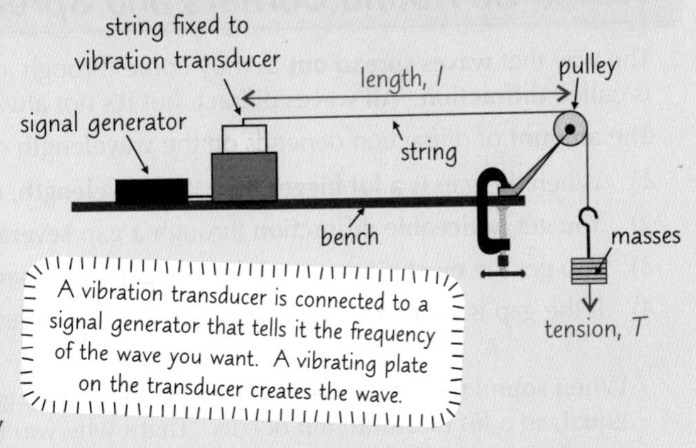

2) Set up the apparatus shown in the diagram with one of your strings. Record μ, measure and record the **length** (l) and work out the **tension** (T) using: $T = mg$ where m is the total mass of the masses in kg

3) Turn on the **signal generator** and vary the frequency until you find the **first harmonic** — i.e. a stationary wave that has a **node** at each end and a single **antinode**. This is the **frequency** of the first harmonic, f.

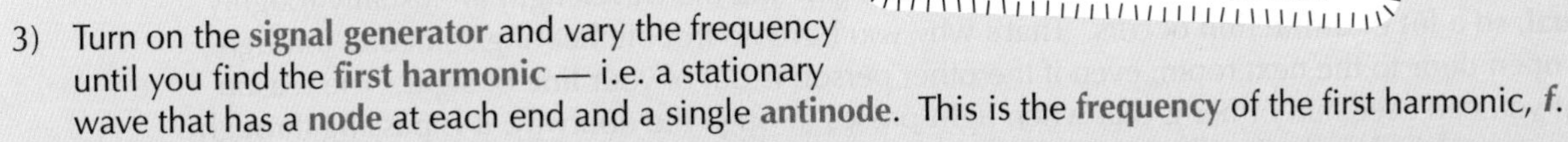

Then investigate how the **length, tension** or **mass per unit length** of the string affects the **resonant frequency** by:

1) Keeping the string **type** (μ) and the **tension** (T) in it the same and altering the **length** (l). Do this by moving the **vibration transducer** towards or away from the pulley. Find the **first harmonic** again, and record f against l.
2) Keeping the string **type** (μ) and **length** (l) the same and **adding** or **removing masses** to change the tension (T). Find the first harmonic again and record f against T.
3) Keeping the **length** (l) and **tension** (T) the same, but using **different string** samples to vary μ. Find the first harmonic and record f against μ.

You can do all of this with a different harmonic — just remember to use the same one throughout the experiment so you're comparing the same resonant frequency.

You should find the following from your investigation:

1) The **longer** the string, the **lower** the resonant frequency — because the **half wavelength** at the resonant frequency is longer.
2) The **heavier** (i.e. the more mass per unit length) the string, the **lower** the resonant frequency — because waves travel more **slowly** down the string. For a given **length** a **lower** wave speed, **c**, makes a **lower** frequency, **f**.
3) The **looser** the string the **lower** the resonant frequency — because waves travel more **slowly** down a **loose** string.

The **frequency** of the first harmonic, f, is: $f = \frac{1}{2l}\sqrt{\frac{T}{\mu}}$ Where l is the string length in m, T is the tension in the string and μ is the mass per unit length of the string.

Practice Questions

Q1 How do stationary waves form?

Q2 At four times the frequency of the first harmonic, how many half wavelengths would fit on a violin string?

Q3 Describe an experiment to investigate stationary waves in a column of air.

Q4 How does the displacement of a particle at one antinode compare to that of a particle at another antinode?

Exam Question

Q1 A stationary wave at the first harmonic frequency, 10 Hz (to 2 s.f.), is formed on a stretched string of length 1.2 m.

a) Calculate the wavelength of the wave. [2 marks]

b) The tension is doubled whilst all other factors remain constant. The frequency is adjusted to once more find the first harmonic of the string. Calculate the new frequency of the first harmonic. [3 marks]

c) Explain how the variation of amplitude along the string differs from that of a progressive wave. [2 marks]

Don't get tied up in knots...

Just remember that two progressive waves can combine to make a stationary wave. How many nodes there are shows what harmonic it is (e.g. 2 nodes = 1st harmonic), and you can change its frequency by changing the medium its in.

Diffraction

This page is essentially about shining light through small gaps and creating pretty patterns. Aaaahh look, a rainbow.

Waves Go **Round Corners** and **Spread** Out of **Gaps**

The way that **waves spread out** as they come through a **narrow gap** or go round obstacles is called **diffraction**. **All** waves diffract, but it's not always easy to observe.

The **amount** of diffraction depends on the **wavelength** of the wave compared with the **size of the gap**.

1) When the gap is **a lot bigger** than the **wavelength**, diffraction is **unnoticeable**.
2) You get **noticeable diffraction** through a gap **several** wavelengths wide.
3) You get the **most** diffraction when the gap is **the same** size as the **wavelength**.
4) If the gap is **smaller** than the wavelength, the waves are mostly just **reflected back**.

When **sound** passes through a **doorway**, the **size of gap** and the **wavelength** are usually roughly **equal**, so **a lot** of **diffraction** occurs. That's why you have no trouble **hearing** someone through an **open door** to the next room, even if the other person is out of your **line of sight**. The reason that you can't **see** him or her is that when **light** passes through the doorway, it is passing through a **gap** around a **hundred million times bigger** than its wavelength — the amount of diffraction is **tiny**. So to get **noticeable** diffraction with light, you must shine it through a very **narrow** slit.

Light Shone Through a **Narrow Slit** Can Form a **Diffraction Pattern**

To observe a clear **diffraction pattern** for light, you need to use a **monochromatic**, **coherent** light source. Monochromatic just means all the light has the same **wavelength** (and frequency) and so is the same **colour**. **Lasers** are a monochromatic and coherent light source.

Demonstrating Light Diffraction Patterns with a Laser

If the wavelength of light is about the **same size** as the aperture, you get a diffraction pattern.

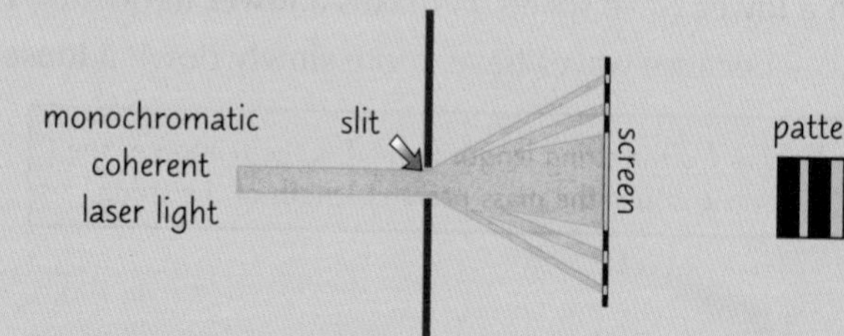

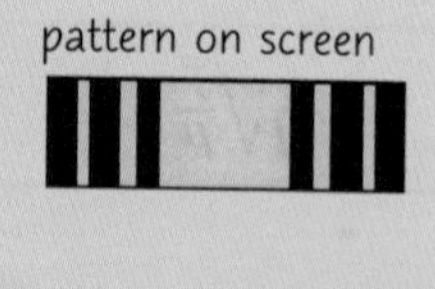

You need to be very careful when using lasers — see page 32.

You'll see a **central bright fringe** (central maximum), with dark and bright fringes **alternating** on either side. The dark and bright fringes are caused by **destructive** and **constructive interference** of light waves (see p.26).

Diffracted **White Light** Creates **Spectra** of Colours

1) **White light** is actually a **mixture** of different colours, each with different **wavelengths**.
2) When white light is shone through a single narrow slit, all of the different wavelengths are **diffracted** by different amounts.
3) This means that **instead** of getting clear **fringes** (as you would with a **monochromatic** light source) you get **spectra** of colours.

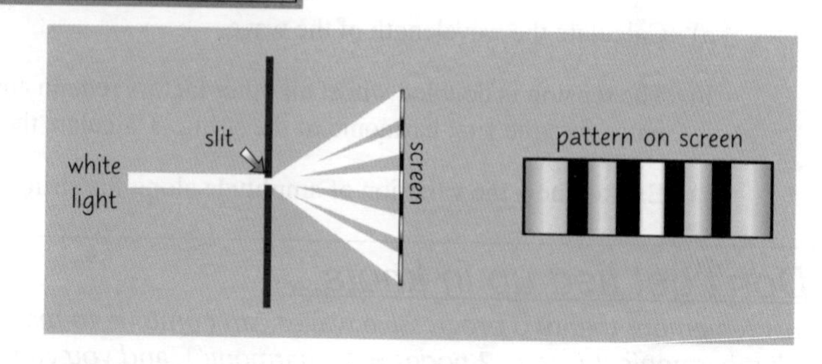

Diffraction

Intensity of Light Means Number of Photons

1) The **central maximum** in a single slit light diffraction pattern is the **brightest** part of the pattern.
2) This is because the **intensity** of light is highest in the centre.
3) **Intensity** is the **power per unit area**.
4) For monochromatic light, all photons have the **same energy**, so an increase in the intensity means an increase in the **number of photons per second**.
5) So there are **more photons** per **unit area** hitting the central maximum per second than the other bright fringes.

Fred wasn't sure about his fringe intensity.

See p.16 for the photon model of light.

The Width of the Central Maximum Varies with Wavelength and Slit Size

When light is shone through a single slit, there are **two** things which affect the **width** of the central maximum:

1) **Increasing** the **slit width** decreases the amount of diffraction. This means the central maximum is narrower, and the **intensity** of the central maximum is **higher**.
2) **Increasing** the **wavelength** increases the amount of diffraction. This means the central maximum is wider, and the intensity of the central maximum is **lower**.

Practice Questions

Q1 What is diffraction?

Q2 Do all waves diffract?

Q3 a) Sketch the pattern produced when monochromatic light is shone through a narrow slit onto a screen.

b) If the wavelength of the monochromatic light was decreased, what would happen to the central maximum?

Q4 What would you expect to see when white light is shone through a thin slit onto a screen?

Q5 What happens to the number of photons in a light beam if the intensity increases?

Exam Questions

Q1 A fire alarm can be heard from the next room through a doorway, even though it is not in the line of sight. Explain why this happens, with reference to the wavelengths of sound and light waves.

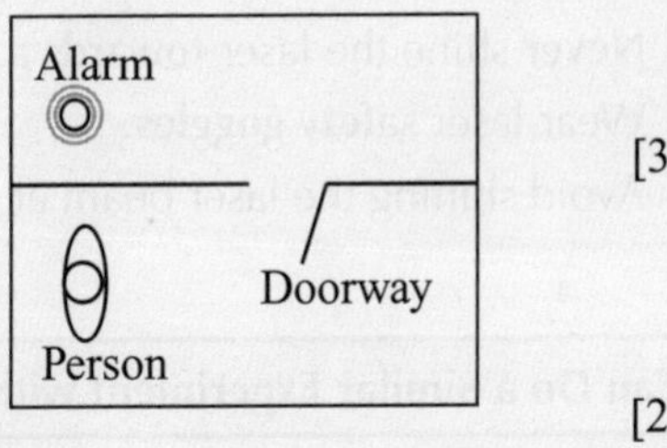

[3 marks]

Q2 A student shines a laser beam through a narrow slit onto a screen.

a) State and explain one reason why using laser light produces a clearer diffraction pattern than other light sources. [2 marks]

b) The student uses a narrower slit. Describe how this will affect the central maximum of the diffraction pattern. Explain your answer. [3 marks]

Waves are just like me at the weekend — they like to spread out...

An important point to remember is that diffraction's only noticeable when the wavelength is roughly equal to the size of the gap the wave is going through. Different light sources give different patterns when diffracted, make sure you know this and can explain why it happens. Diffraction crops up again in physics — so make sure you really understand it.

Two-Source Interference

You know what happens with one light source, so now it's time to see what happens with two. I can hardly wait...

Demonstrating **Two-Source** Interference in **Water** and **Sound** is Easy

1) It's **easy** to demonstrate **two-source interference** for either **sound** or **water** because they've got **wavelengths** of a handy **size** that you can **measure**.
2) You need **coherent** sources, which means the **wavelength** and **frequency** have to be the **same**. The trick is to use the **same oscillator** to drive **both sources**. For **water**, one **vibrator drives two dippers**. For sound, **one oscillator** is connected to **two loudspeakers**. (See diagram on page 27.)

Demonstrating **Two-Source** Interference for **Light** is Harder

Young's Double-Slit Experiment

1) To see **two-source** interference with light, you can either use two separate, **coherent** light sources or you can shine a **laser** through **two slits**. Laser light is **coherent** and **monochromatic**.
2) Young's double-slit experiment shines a laser through two slits onto a **screen**.
3) The slits have to be about the **same size** as the **wavelength** of the laser light so that it is diffracted — then the light from the slits acts like two coherent point sources.

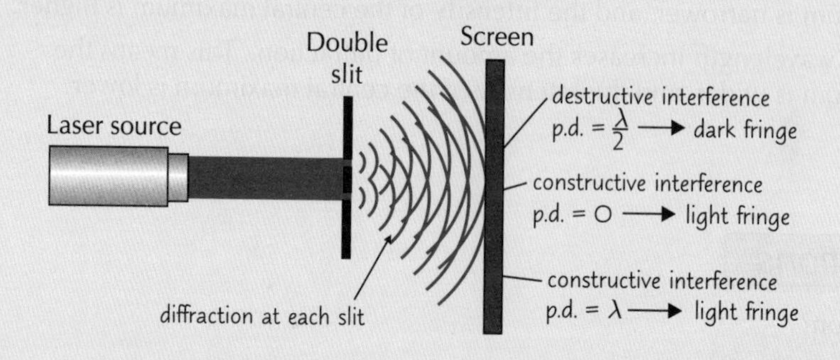

4) You get a pattern of light and dark **fringes**, depending on whether constructive or destructive **interference** is taking place. Thomas Young — the first person to do this experiment (with a lamp rather than a laser) — came up with an **equation** to **work out** the **wavelength** of the **light** from this experiment (see p.33).

Working with lasers is very **dangerous** because laser light is focused into a very direct, powerful beam of monochromatic light. If you looked at a laser beam **directly**, your eye's lens would focus it onto your retina, which would be **permanently damaged**.

To make sure you don't cause **damage** while using lasers, you should:

1) **Never** shine the laser **towards** a person.
2) Wear laser **safety goggles**.
3) Avoid shining the laser beam at a **reflective surface**.
4) Have a **warning sign** on display.
5) Turn the laser **off** when it's not needed.

You Can Do a Similar Experiment with Microwaves

To see interference patterns with **microwaves**, you can **replace** the laser and slits with two microwave **transmitter cones** attached to the **same** signal generator.

You also need to replace the screen with a microwave **receiver probe** (like the one used in the stationary waves experiment on page 28).

If you move the probe along the path of the green arrow, you'll get an **alternating pattern** of **strong** and **weak** signals — just like the light and dark fringes on the screen.

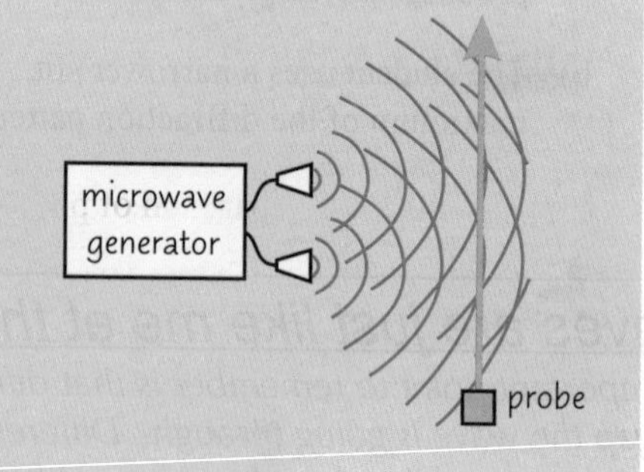

Two-Source Interference

Work Out the Wavelength with **Young's Double-Slit Formula**

1) The fringe spacing (**w**), wavelength (**λ**), spacing between slits (**s**) and the distance from slits to screen (**D**) are all related by **Young's double-slit formula**, which works for all waves.

Fringe spacing, $w = \frac{\lambda D}{s}$

'Fringe spacing' means the distance from the centre of one minimum to the centre of the next minimum or from the centre of one maximum to the centre of the next maximum.

Always check your fringe spacing.

2) Since the wavelength of light is so small you can see from the formula that a high ratio of ***D* / *s*** is needed to make the fringe spacing **big enough to see**.
3) Rearranging, you can use **$\lambda = ws / D$** to **calculate the wavelength** of light.
4) The fringes are **so tiny** that it's very hard to get an **accurate value of *w***. It's easier to measure across **several** fringes then **divide** by the number of **fringe widths** between them.

Young's Experiment was **Evidence** for the **Wave Nature** of EM Radiation

1) Towards the end of the **17th century**, two important **theories of light** were published — one by Isaac Newton and the other by a chap called Huygens. **Newton's** theory suggested that light was made up of tiny particles, which he called "**corpuscles**". And **Huygens** put forward a theory using **waves**.
2) The **corpuscular theory** could explain **reflection** and **refraction**, but **diffraction** and **interference** are both **uniquely** wave properties. If it could be **shown** that light showed interference patterns, that would help settle the argument once and for all.
3) **Young's** double-slit experiment (over 100 years later) provided the necessary evidence. It showed that light could both **diffract** (through the narrow slits) and **interfere** (to form the interference pattern on the screen).
Of course, this being Physics, nothing's ever simple — give it another 100 years or so and the debate would be raging again, (p.16).

Practice Questions

Q1 In Young's experiment, why do you get a bright fringe at a point equidistant from both slits?
Q2 What does Young's experiment show about the nature of light?
Q3 Write down Young's double-slit formula.

Exam Questions

Q1 a) The diagram on the right shows two coherent light sources, S_1 and S_2, being shone on a screen. State the measurements you would need to take to calculate the wavelength of the light. [3 marks]

b) S_1 and S_2 may be slits in a screen behind which there is a source of laser light, instead of being two separate sources. State two safety precautions you should take when using this set-up. Explain your answer. [2 marks]

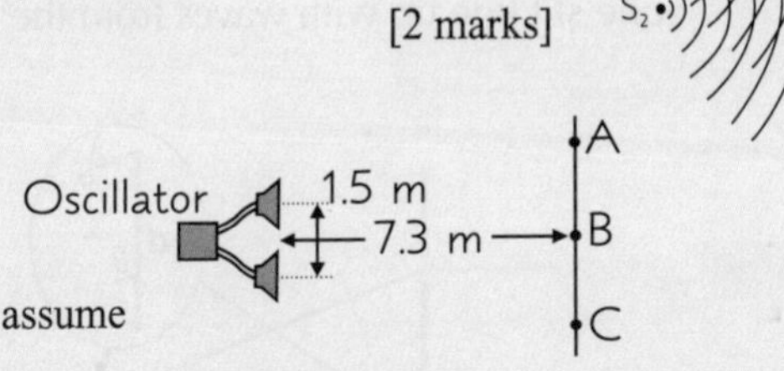

Q2 In an experiment to study sound interference, two loudspeakers are connected to an oscillator emitting sound at 1320 Hz and set up as shown in the diagram. They are 1.5 m apart and 7.3 m away from the line AC. A listener moving along the line hears minimum sound at A, maximum sound at B and minimum sound again at C. (You may assume that Young's double-slit formula can be used in this calculation).

a) Calculate the wavelength of the sound waves if the speed of sound in air is taken to be 330 ms^{-1}. [1 mark]

b) Calculate the separation of points A and C. [1 mark]

Learn this stuff — or you'll be playing ketchup...

A few things to learn here — some diffraction experiments, a formula and a little bit of history. Be careful when you're calculating the fringe width by averaging over several fringes. Don't just divide by the number of bright lines. Ten bright lines will only have nine fringe widths between them, not ten. It's an easy mistake to make, but you have been warned.

Diffraction Gratings

Ay... starting to get into some pretty funky stuff now. I like light experiments.

Interference Patterns Get **Sharper** When You Diffract Through **More Slits**

1) You can repeat **Young's double-slit** experiment (see p.32) with **more than two equally spaced** slits. You get basically the **same shaped** pattern as for two slits — but the **bright bands** are **brighter** and **narrower** and the **dark areas** between are **darker**.
2) When **monochromatic light** (one wavelength) is passed through a **grating** with **hundreds** of slits per millimetre, the interference pattern is **really sharp** because there are so **many beams reinforcing** the **pattern**.
3) Sharper fringes make for more **accurate** measurements.

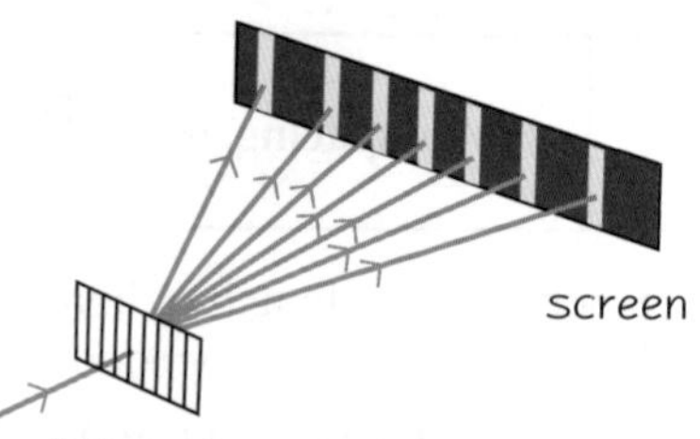

Monochromatic Light on a **Diffraction Grating** gives **Sharp Lines**

1) For **monochromatic** light, all the **maxima** are sharp lines. (It's different for white light — see page 30.)
2) There's a line of **maximum brightness** at the centre called the **zero order** line.
3) The lines just **either side** of the central one are called **first order lines**. The **next pair out** are called **second order** lines and so on.
4) For a grating with slits a distance d apart, the angle between the **incident beam** and **the nth order maximum** is given by:

$$d \sin \theta = n\lambda$$

5) So by observing d, θ and n you can **calculate the wavelength** of the light.

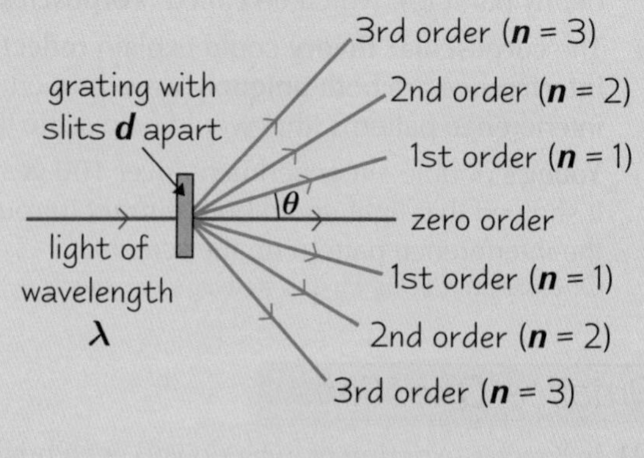

If the grating has N slits per metre, then the slit spacing, d, is just $1/N$ metres.

DERIVING THE EQUATION:

1) At **each slit**, the incoming waves are **diffracted**. These diffracted waves then **interfere** with each other to produce an **interference pattern**.
2) Consider the **first order maximum**. This happens at the **angle** when the waves from one slit line up with waves from the **next slit** that are **exactly one wavelength** behind.

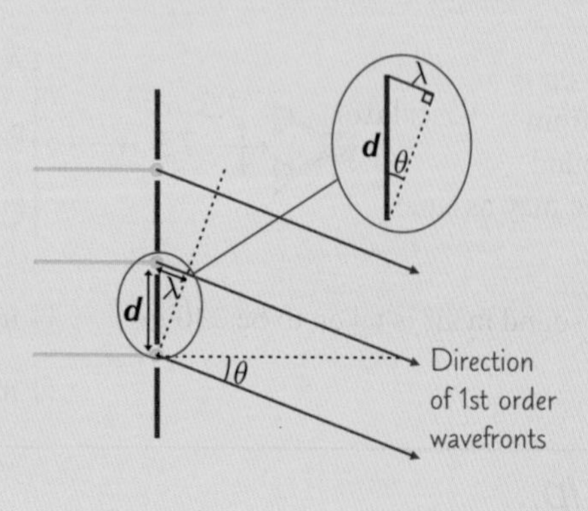

3) Call the **angle** between the **first order maximum** and the **incoming light** θ.
4) Now, look at the **triangle** highlighted in the diagram. The angle is θ (using basic geometry), d is the slit spacing and the **path difference** is λ.
5) So, for the first maximum, using trig:
$$d \sin \theta = \lambda$$
6) The other maxima occur when the path difference is 2λ, 3λ, 4λ, etc. So to make the equation **general**, just replace λ with $n\lambda$, where n is an integer — the **order** of the maximum.

Diffraction Gratings

You can Draw General Conclusions from d sin θ = nλ

1) If λ is **bigger, sin θ** is **bigger**, and so θ is **bigger**. This means that the larger the **wavelength**, the more the pattern will **spread out**.
2) If d is **bigger, sin θ is smaller**. This means that the **coarser** the **grating**, the **less** the pattern will **spread out**.
3) Values of **sin θ** greater than **1** are **impossible**. So if for a certain n you get a result of **more than 1** for **sin θ** you know that that order **doesn't exist**.

Diffraction Gratings Help to Identify Elements and Calculate Atomic Spacing

1) **White light** is really a **mixture** of **colours**. If you **diffract** white light through a **grating** then the patterns due to **different wavelengths** within the white light are **spread out** by **different** amounts.
2) Each **order** in the pattern becomes a **spectrum**, with **red** on the **outside** and **violet** on the **inside**. The **zero order maximum** stays **white** because all the wavelengths just pass straight through.
3) **Astronomers** and **chemists** often need to study spectra to help identify elements. They use diffraction gratings rather than prisms because they're **more accurate**.

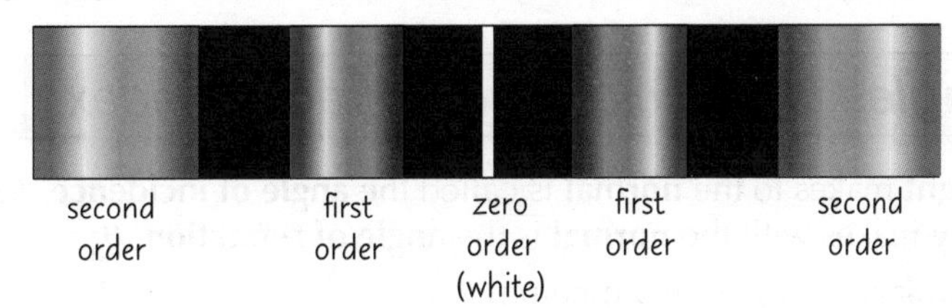

4) The wavelength of **X-rays** is of a similar scale to the spacing between **atoms** in crystalline solids. This means that X-rays will form a **diffraction pattern** when directed at a thin crystal.
5) The crystal acts like a **diffraction grating** and the spacing between **atoms** (slit width) can be found from the diffraction pattern.
6) This is called **X-ray crystallography** — it was used to discover the structure of **DNA**.

Practice Questions

Q1 How is the diffraction grating pattern for white light different from the pattern for laser light?

Q2 What difference does it make to the pattern if you use a finer grating?

Q3 What equation is used to find the angle between the *n*th order maximum and the incident beam for a diffraction grating?

Q4 Derive the equation you quoted in Q3.

Exam Questions

Q1 Yellow laser light of wavelength 6.0×10^{-7} m is transmitted through a diffraction grating of 4.0×10^{5} lines per metre.

a) Calculate the angle to the normal at which the first and second order bright lines are seen. [3 marks]

b) State whether there is a fifth order line. Explain your answer. [1 mark]

Q2 Visible, monochromatic light is transmitted through a diffraction grating of 3.7×10^{5} lines per metre. The first order maximum is at an angle of 14.2° to the incident beam. Calculate the wavelength of the incident light. [2 marks]

Ooooooooooooo — pretty patterns...

Derivation — ouch. At least it's not a bad one though. As long as you learn the diagram, it's just geometry and a bit of trig from there. Make sure you learn the equation — that way, you know what you're aiming for. As for the rest of the page, remember that the more slits you have, the sharper the image — and white light makes a pretty spectrum.

Refractive Index

The stuff on the next two pages explains why your legs look short in a swimming pool.

*The **Refractive Index** of a Material Measures **How Much** It Slows Down Light*

Light goes fastest in a **vacuum**. It **slows down** in other materials, because it **interacts** with the particles in them. The more **optically dense** a material is, the more light slows down when it enters it.

The **absolute refractive index** of a material, ***n***, is a measure of **optical density**. It is found from the **ratio** between the **speed of light** in a **vacuum**, **c**, and the speed of light in that **material**, c_s.

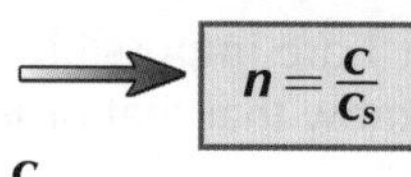

$$n = \frac{c}{c_s}$$

$c = 3.00 \times 10^8$ ms^{-1}

The **relative** refractive index **between two materials**, $_1n_2$, is the ratio of the speed of light **in material 1** to the speed of light **in material 2**.

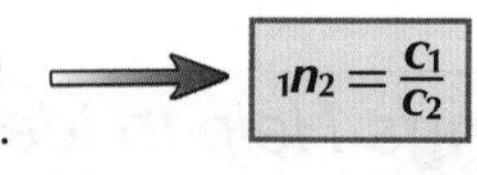

$$_1n_2 = \frac{c_1}{c_2}$$

The speed of light in air is only a tiny bit smaller than c. So you can assume the refractive index of air is 1.

Combining the two equations gives: $_1n_2 = \frac{n_2}{n_1}$

1) The **absolute refractive index** of a material is a **property** of that material only. A **relative refractive index** is a property of the **interface** between two materials. It's different for **every possible pair**.
2) Because you can assume $n_{air} = 1$, you can assume the refractive index for an **air to glass boundary** equals the **absolute refractive index** of the glass.

***Snell's Law** uses **Angles** to Calculate the Refractive Index*

1) The **angle** the **incoming light** makes to the **normal** is called the **angle of incidence**, θ_1. The **angle** the **refracted ray** makes with the **normal** is the **angle of refraction**, θ_2.
2) The light is crossing a **boundary**, going from a medium with **refractive index** n_1 to a medium with refractive index n_2.
3) When light enters an **optically denser** medium it is refracted **towards** the normal.
4) n, θ_1 and θ_2 are related by **Snell's law**: $n_1 \sin\theta_1 = n_2 \sin\theta_2$

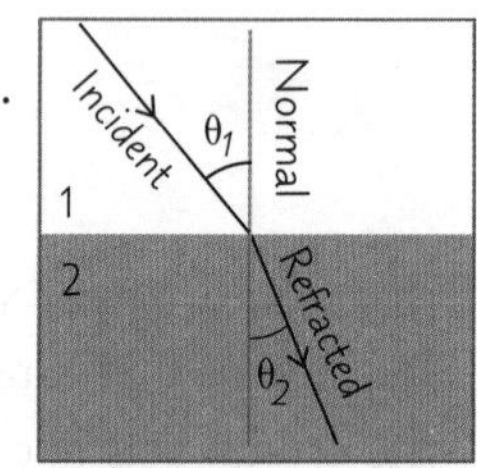

*Light Leaving an **Optically Denser Material** is Refracted **Away** from the **Normal***

When light **goes from** an optically denser material into an optically **less dense** material (e.g. glass to air), interesting things can happen.

Refracted
n_2
n_1
Reflected
θc
Incident

1) Shine a ray of light at a boundary going from refractive index n_1 to n_2, then gradually **increase** the angle of incidence.
2) The light is refracted away from the normal, so as you increase the angle of incidence, the angle of **refraction** gets closer and closer to **90°**.
3) Eventually θ_1 reaches a **critical angle** θ_c for which $\theta_2 = 90°$. The light is **refracted** along the **boundary**.
4) As sin 90° = 1, Snell's law, $n_1 \sin\theta_1 = n_2 \sin\theta_2$, becomes $n_1 \sin\theta_c = n_2 \times 1$ so:

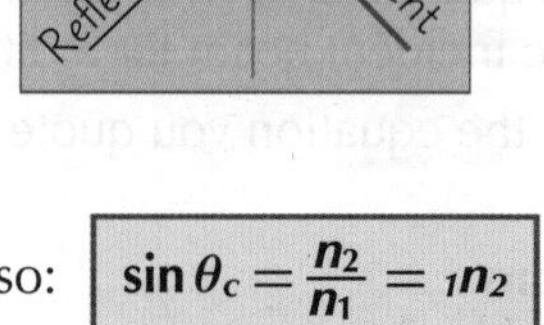

$$\sin\theta_c = \frac{n_2}{n_1} = {_1n_2}$$

5) At θ_1 **greater** than the **critical angle**, refraction is impossible. All the light is **reflected** back into the material — this is called **total internal reflection**.

Since sin can only take values between -1 and 1, total internal reflection can only happen if $\sin\theta_c < 1$ so $_1n_2 < 1$.

Optical Fibres** Use **Total Internal Reflection

An optical fibre is a very **thin flexible tube** of **glass** or **plastic** fibre that can carry **light signals** over long distances and round corners. You only need to know about **step-index** optical fibres.

Cladding
Light
Optical fibre

1) Step-index optical fibres themselves have a **high refractive index** but are surrounded by **cladding** with a lower refractive index to allow **total internal reflection**. Cladding also protects the fibre from **scratches** which could let **light escape**.
2) Light is shone in at **one end** of the fibre. The fibre is so **narrow** that the light always **hits the boundary** between the fibre and cladding at an **angle bigger** than the **critical angle**.
3) So all the light is **totally internally reflected** from boundary to boundary until it reaches the other end.

Refractive Index

Dispersion and Absorption Cause Signal Degradation

A **signal** (a stream of pulses of light) travelling down an optical fibre can be **degraded** by **absorption** or by **dispersion**. **Signal degradation** can cause **information** to be **lost**.

Absorption Causes Loss in Amplitude

As the signal travels, some of its energy is lost through **absorption** by the **material** the fibre is made from. This energy loss results in the **amplitude** of the signal being **reduced**.

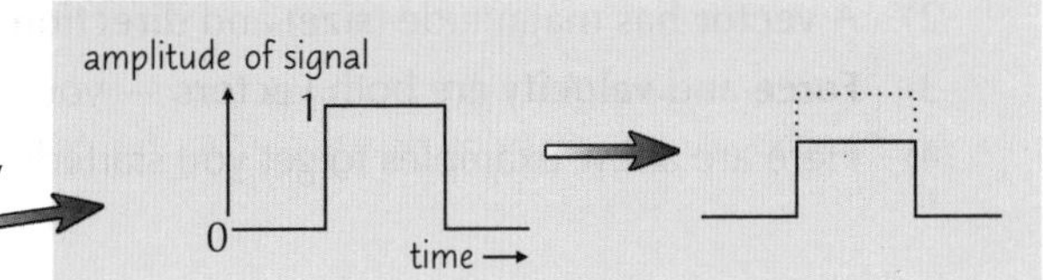

Dispersion Causes Pulse Broadening

There are two types of dispersion that can degrade a signal.

1) **Modal dispersion** — light rays enter the fibre at different angles, and so take different **paths**. The rays which take a **longer path** take longer to reach the other end than those that travel down the **middle** of the fibre. A **single-mode** fibre only lets light take **one path**, so it stops modal dispersion.
2) **Material dispersion** — light consists of different **wavelengths** that travel at different speeds in the fibre — this causes some light wavelengths to reach the end of the fibre faster than others. Using **monochromatic light** can stop material dispersion.

Both types of dispersion lead to **pulse broadening**.
The signal sent down the fibre is broader at the other end.
Broadened pulses can **overlap** each other and confuse the signal.

An **optical fibre repeater** can be used to **boost** and regenerate the signal every so often, which can **reduce** signal **degradation** caused by both **absorption** and **dispersion**.

Practice Questions

Q1 Why does light go fastest in a vacuum and slow down in other media?

Q2 What is the formula for the critical angle for a ray of light at a water/air boundary?

Exam Questions

Q1 a) Light travels in diamond at 1.24×10^8 ms^{-1}. Calculate the refractive index of diamond. [1 mark]

b) Calculate the angle of refraction if light strikes a facet of a diamond ring at an angle of 50° (to 2 s.f.) to the normal of the air/diamond boundary. [2 marks]

Q2 An adjustable underwater spotlight is placed on the floor of an aquarium tank. When the light points upwards at a steep angle a beam comes through the surface of the water into the air, and the tank is dimly lit. When the spotlight is placed at a shallower angle, no light comes up through the water surface, and the tank is brightly lit.

a) Explain what is happening. [2 marks]

b) It is found that the beam into the air disappears when the spotlight is pointed at any angle of less than 41.25° to the floor. Calculate the refractive index of water. [2 marks]

Q3 a) Explain the ways in which the cladding is designed to keep transmitted light inside an optical fibre. [2 marks]

b) The cladding functions as expected, but there is still some information loss when a step-index optical fibre is used to transmit light signals over long distances.
Discuss the potential causes of this loss of information and how the design and operation of the optical fibre could be altered to reduce information loss over long transmission distances. [6 marks]

I don't care about expensive things — all I care about is wave speed...

Physics examiners are always saying how candidates do worst in the waves bit of the exam. You'd think they'd have something more important to worry about — third world poverty, war, Posh & Becks... But no.

Scalars and Vectors

Mechanics is one of those things that you either love or hate. I won't tell you which side of the fence I'm on.

Scalars** Only Have **Size**, but **Vectors** Have **Size** and **Direction

1) A **scalar** has **no direction** — it's **just an amount** of something, like the **mass** of a **sack of meaty dog food**.
2) A **vector** has magnitude (**size**) and **direction** — like the **speed and direction** of next door's **cat** running away.
3) **Force** and **velocity** are both **vectors** — you need to know **which way** they're going as well as **how big** they are.
4) Here are a few examples to get you started:

Scalars	Vectors
mass, temperature, time, length/distance, speed, energy	displacement, velocity, force (including weight), acceleration, momentum

*There are **Two Methods** for **Adding Vectors** Together*

Adding two or more vectors is called finding the **resultant** of them.
There are two ways of doing this you need to know about.

Scale Drawings

Start by making a **scale drawing** of the two vectors (tip-to-tail if they're not already), draw the **resultant vector** from the tail of the first to the tip of the last, and measure its **length** and **angle**.

Example: A man walks 3.0 m on a bearing of 055° then 4.0 m east. Find the magnitude and direction (to the nearest degree) of his displacement, s.

The man's 'displacement' gives his position relative to his starting point.

Start by drawing a **scale diagram** for how far the man walked using a **ruler** and a **protractor**.

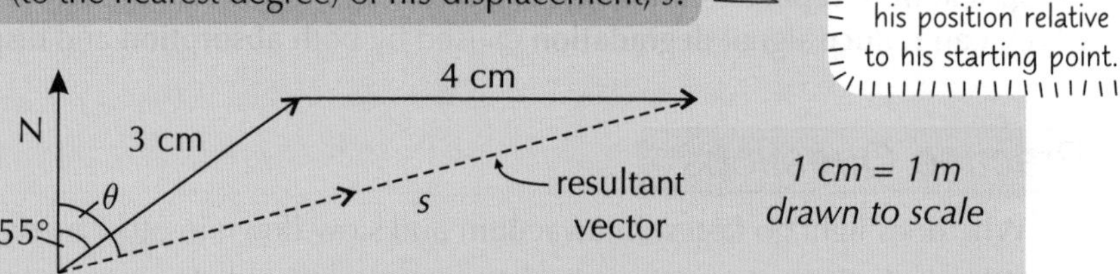

Then just **measure** the missing side with a ruler and the missing angle with a protractor:
s = 6.7 cm and θ = 75° (to the nearest degree)
So the man's displacement is **6.7 m**, on a bearing of **075°**.

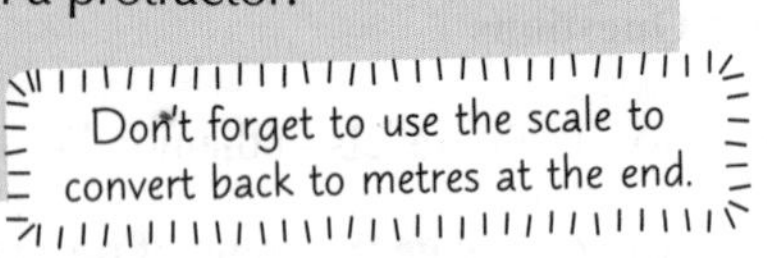

Pythagoras** and **Trigonometry

When two vectors are at **right angles** to each other, you can use maths to work it out without a scale drawing.

Example: Jemima goes for a walk. She walks 3 m north and 4 m east. She has walked 7 m but she isn't 7 m from her starting point. Find the magnitude and direction (to the nearest degree) of her displacement.

First, sketch the vectors **tip-to-tail**. Then draw a line from the **tail** of the first vector to the **tip** of the last vector to give the **resultant**:
Because the vectors are at right angles, you get the **magnitude** of the resultant using Pythagoras:

4 m
3 m
resultant vector, R

Jemima

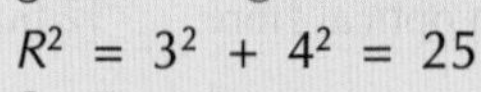

$R^2 = 3^2 + 4^2 = 25$
So R = **5 m**

Now find the **bearing** of Jemima's new position from her original position.
You use the triangle again, but this time you need to use trigonometry. You know the opposite and the adjacent sides, so you need to use:
$\tan\theta = 4 / 3$
θ = **053°**

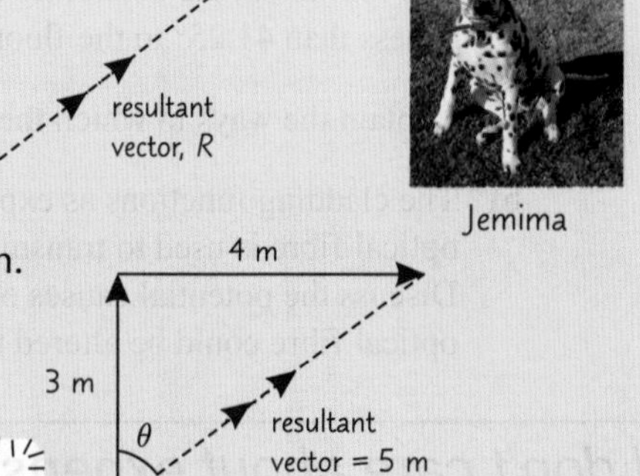

Trig's really useful in mechanics — so make sure you're completely okay with it. Remember SOH CAH TOA.

Scalars and Vectors

Sometimes you have to do it backwards.

*It's Useful to Split a **Vector** into **Horizontal** and **Vertical Components***

This is the opposite of finding the resultant — you start from the resultant vector and split it into two **components** at right angles to each other. You're basically **working backwards** from the examples on the other page.

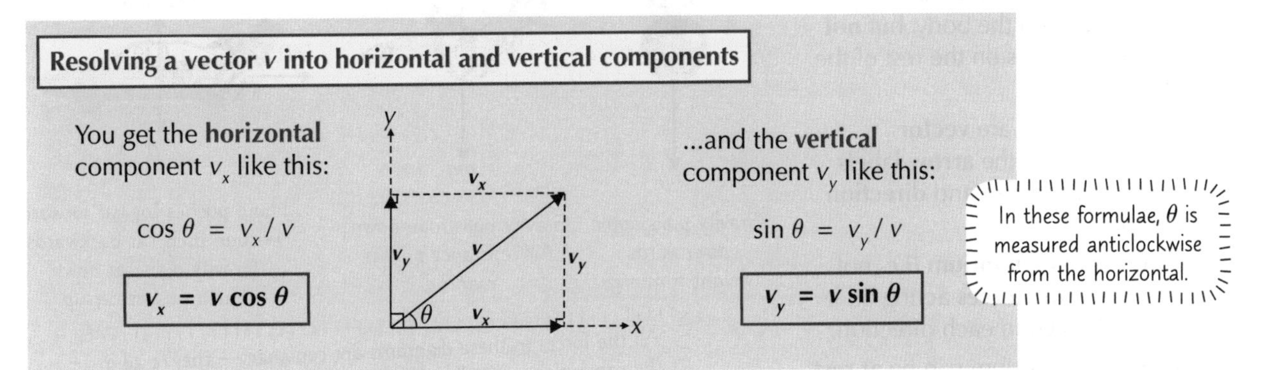

Example: Charley's amazing floating home is travelling at a speed of 5.0 ms^{-1} at an angle of 60° (to 2 s.f.) up from the horizontal. Find the vertical and horizontal components.

The **horizontal** component v_x is:

$v_x = v\cos\theta = 5\cos 60° = \mathbf{2.5\ ms^{-1}}$

The vertical component v_y is:

$v_y = v\sin\theta = 5\sin 60° = \mathbf{4.3\ ms^{-1}}$ **(to 2 s.f.)**

4.3 ms^{-1} 5 ms^{-1} 60° 2.5 ms^{-1}

Charley's mobile home was the envy of all his friends.

Resolving is dead useful because the two components of a vector **don't affect each other**. This means you can deal with the two directions **completely separately**.

vertical motion — horizontal motion

Only the vertical component is affected by gravity.

Practice Questions

Q1 Explain the difference between a scalar quantity and a vector quantity.

Q2 Jemima is chasing a mechanised rabbit. She follows it for 50 m in a south-east direction. It then changes direction, and she follows it a further 80 m west before catching it. By drawing a scale diagram, show that when Jemima catches the rabbit her displacement is 57 m on a bearing of 218° (to the nearest degree).

Q3 Jemima has gone for a swim in a river which is flowing at 0.35 ms^{-1}. She swims at 0.18 ms^{-1} at right angles to the current. Show that her resultant velocity is 0.39 ms^{-1} (to 2 s.f.) at an angle of 27° (to 2 s.f.) to the current.

Exam Questions

Q1 The wind is creating a horizontal force of 20.0 N on a falling rock of weight 75 N. Calculate the magnitude and direction of the resultant force. [2 marks]

Q2 A glider is travelling at a velocity of 20 ms^{-1} (to 2 s.f.) at an angle of 15° below the horizontal. Calculate the horizontal and vertical components of the glider's velocity. [2 marks]

Mum said my life was lacking direction, so I became a vector collector...

I always think that the hardest part of vector questions is getting the initial diagram right. Once you've worked out what's going on, they're all the same — they're a piece of cake taking a walk in the park. Easy as pie in a light breeze.

Forces

Remember the vector stuff from the last two pages... good, you're going to need it...

Free-Body Force Diagrams Show All Forces on a Single Body

1) **Free-body force** diagrams show a **single body** on its own.
2) The diagram should include all the **forces** that **act on** the body, but **not** the **forces it exerts** on the rest of the world.
3) Remember **forces** are **vector quantities** and so the **arrow labels** should show the **size** and **direction** of the forces.
4) If a body is in **equilibrium** (i.e. not accelerating) the **forces** acting on it will be **balanced** in each direction.
5) A body in equilibrium can be **at rest** or moving with a **constant velocity**.

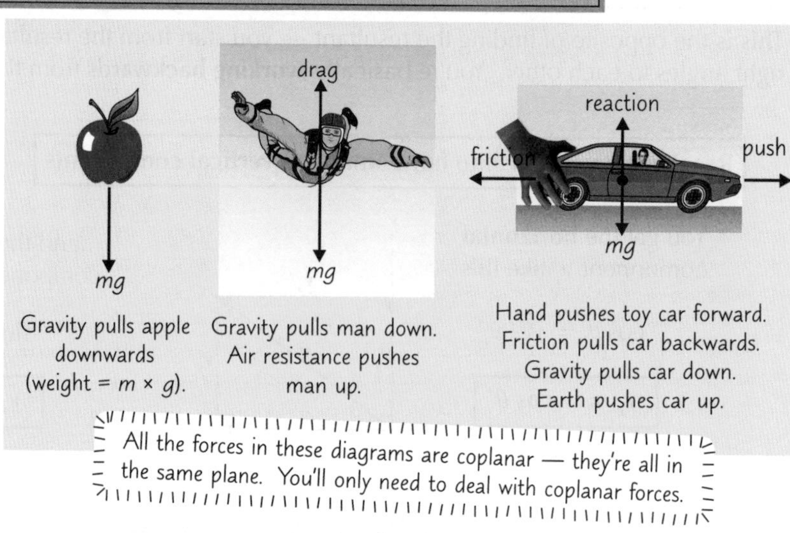

Gravity pulls apple downwards (weight = $m \times g$).

Gravity pulls man down. Air resistance pushes man up.

Hand pushes toy car forward. Friction pulls car backwards. Gravity pulls car down. Earth pushes car up.

All the forces in these diagrams are coplanar — they're all in the same plane. You'll only need to deal with coplanar forces.

Resolving a Force Means Splitting it into Components

1) Forces can be in **any direction**, so they're not always at right angles to each other. This is sometimes a bit **awkward** for **calculations**.
2) To make an 'awkward' force easier to deal with, you can think of it as **two separate forces**, acting at **right angles** to **each other**. Forces are **vectors**, so you can use the same method as on the previous page.

Example: The force F has exactly the same effect as the horizontal and vertical forces, F_H and F_V. Replacing F with F_H and F_V is called **resolving the force F.**

Use these formulas when resolving forces: $\frac{F_H}{F} = \cos\theta$ or $F_H = F\cos\theta$ and $\frac{F_V}{F} = \sin\theta$ or $F_V = F\sin\theta$

Things Get Tricky with 3 Forces

1) When you have **three** coplanar forces acting on a body in **equilibrium**, you can draw the forces as a triangle, forming a **closed loop** like these:
2) Be careful when you draw the triangles not to go into autopilot and draw F_3 as the sum of F_1 and F_2 — it has to be in the **opposite** direction to balance the other two forces.
3) If it's a right-angled triangle, you can use **Pythagoras** to find a missing force.
4) If not, you might have to **resolve the forces** in each direction.

Example: 3 teams are taking part in the CGP Fun Day 3-Way Tug of War. If the knot is in equilibrium, find the size of force P.

P has no vertical component, so you can ignore the vertical components of the other teams (as they must cancel each other out). Resolve the horizontal forces (take right as the positive direction):

$(\cos 41° \times 1175) + (\cos 41° \times 1175) - P = 0$

$P = 886.7... + 886.7... =$ **1800 N (to 2 s.f.)**

You can check your answer by drawing a triangle with the three forces and seeing if it forms a closed loop:

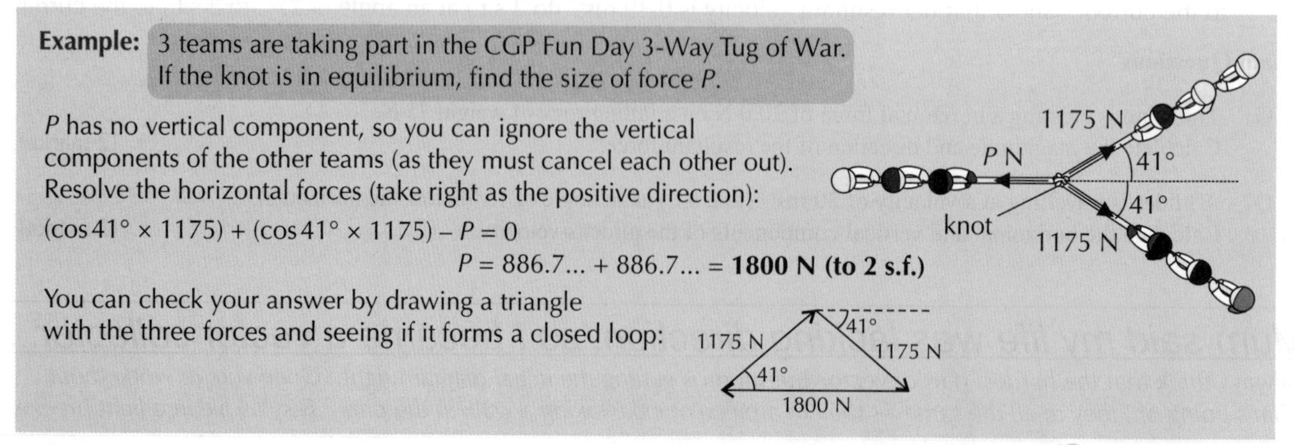

Forces

You Add Components Back Together to get the Resultant Force

1) If **two forces** act on an object, you find the **resultant** (total) **force** by adding the **vectors** together and creating a **closed triangle**, with the resultant force represented by the **third side**.
2) Forces are vectors (as you know), so you use **vector addition** — draw the forces as vector arrows put 'tip-to-tail'.
3) Then it's yet more trigonometry to find the **angle** and the **length** of the third side.

Example: Two dung beetles roll a dung ball along the ground at constant velocity. Beetle A applies a force of 0.50 N northwards while beetle B exerts a force of only 0.20 N eastwards. What is the resultant force on the dung ball?

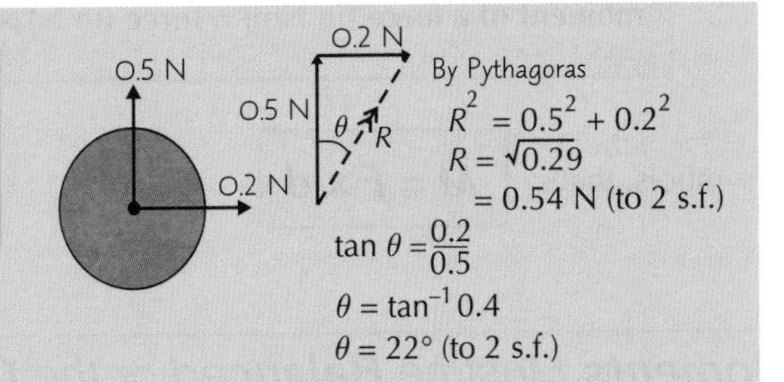

The resultant force is **0.54 N (to 2 s.f.)** at an angle of **22° (to 2 s.f.)** from north.

Choose Sensible Axes for Resolving

Use directions that **make sense** for the situation you're dealing with. If you've got an object on a slope, choose your directions **along the slope** and **at right angles to it**. You can turn the paper to an angle if that helps.

Always choose sensible axes

Examiners like to call a slope an "inclined plane".

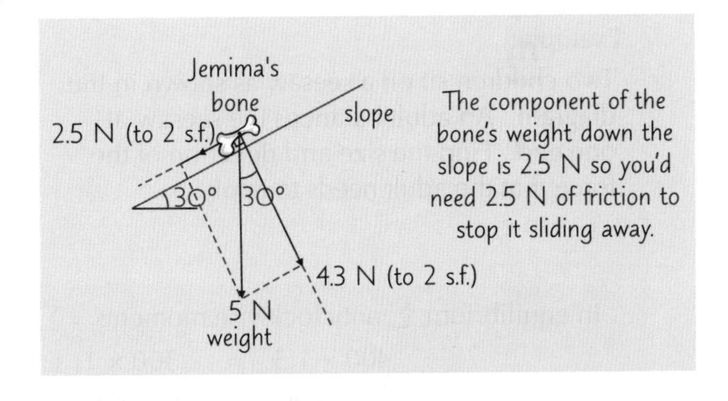

Practice Questions

Q1 Sketch a free-body force diagram for an ice hockey puck moving across the ice (assuming no friction).

Q2 What are the horizontal and vertical components of the force F?

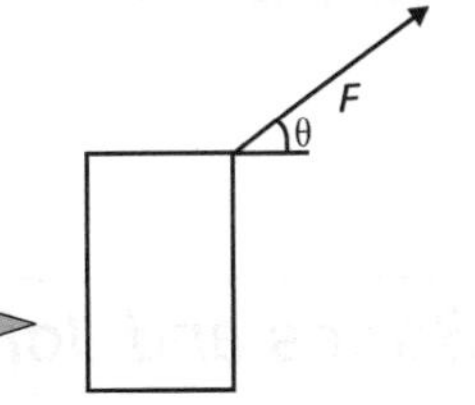

Exam Questions

Q1 A picture with a mass of 8.0 kg is suspended from a hook as shown in the diagram. Calculate the tension force, T, in the string. Use $g = 9.81$ ms^{-2}.

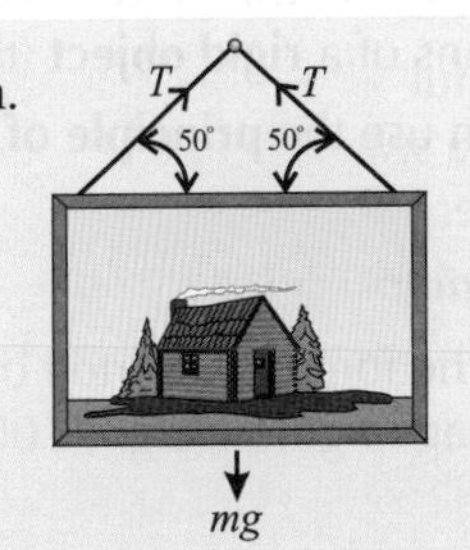

[2 marks]

Q2 Two elephants pull a tree trunk as shown in the diagram. Calculate the resultant force on the tree trunk.

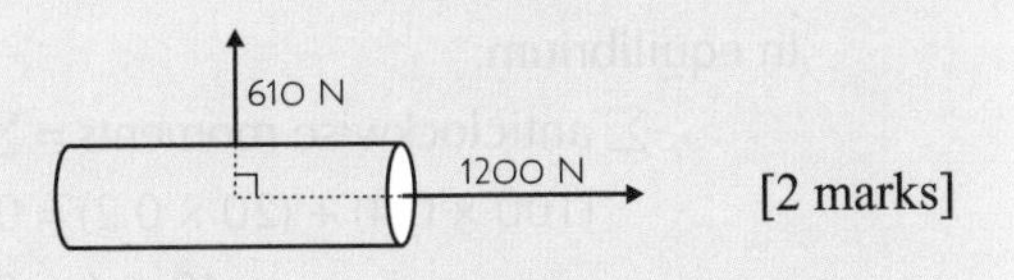

[2 marks]

Free-body force diagram — sounds like it comes with a dance mat...

Remember those $F\cos\theta$ and $F\sin\theta$ bits. Write them on bits of paper and stick them to your wall. Scrawl them on your pillow. Tattoo them on your brain. Whatever it takes — you just have to learn them.

Moments

*This is not a time for jokes. There is not a moment to lose. Oh ho ho ho ho *bang*. (Ow.)*

A Moment is the Turning Effect of a Force

The **moment** of a **force** depends on the **size** of the force and **how far** the force is applied from the **turning point**:

moment of a force (in Nm) = **force** (in N) × **perpendicular distance from the point to the line of action of the force** (in m)

The line of action of a force is a line along which it acts.

In symbols, that's: $M = F \times d$

Moments Must be Balanced or the Object will Turn

The **principle of moments** states that for a body to be in **equilibrium**, the **sum of the clockwise moments** about any point **equals** the **sum of the anticlockwise moments** about the same point.

Example:

Two children sit on a seesaw as shown in the diagram. An adult balances the seesaw at one end. Find the size and direction of the force that the adult needs to apply.

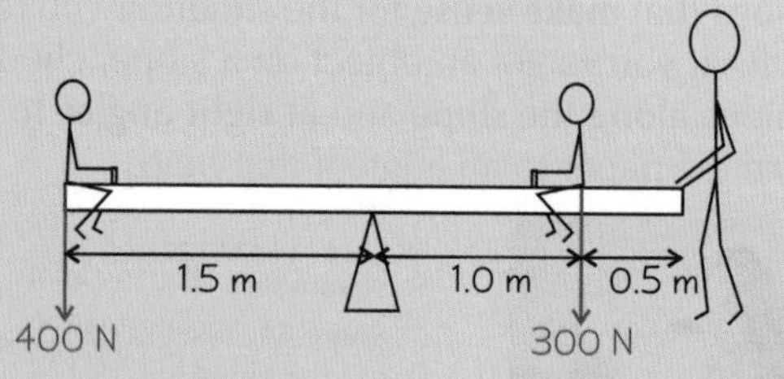

In equilibrium, $\sum$ anticlockwise moments = $\sum$ clockwise moments

$400 \times 1.5 = (300 \times 1) + 1.5F$

$600 = 300 + 1.5F$

Final answer: $F = 200$ N downwards

$\sum$ means "the sum of"

Muscles, Bones and Joints Act as Levers

1) In a lever, an **effort force** (in this case from a muscle) acts against a **load force** (e.g. the weight of your arm) by means of a **rigid object** (the bone) rotating around a **pivot** (the joint).
2) You can use the **principle of moments** to answer lever questions:

Example:

Find the force, E, exerted by the biceps in holding a bag of gold still. The bag of gold weighs 100 N and the forearm weighs 20 N.

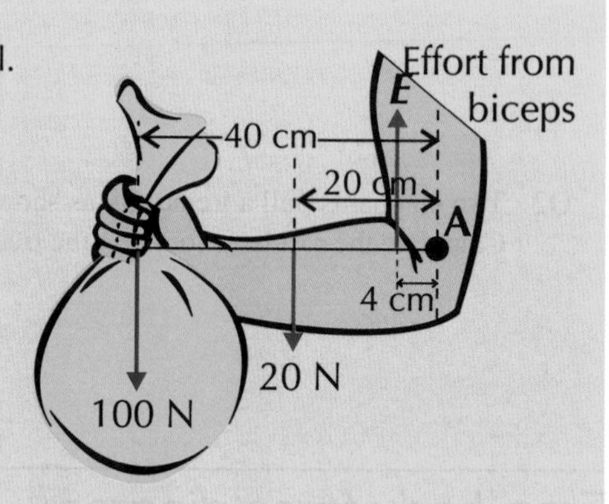

Take moments about **A**.

In equilibrium:

$\sum$ anticlockwise moments = $\sum$ clockwise moments

$(100 \times 0.4) + (20 \times 0.2) = 0.04E$

$40 + 4 = 0.04E$

Final answer: $E = 1100$ N

Moments

A Couple is a Pair of Coplanar Forces

1) A couple is a **pair** of **forces** of **equal size** which act **parallel** to each other, but in **opposite directions.** The forces are **coplanar** (see page 40).
2) A couple doesn't cause any resultant linear force, but **does** produce a **turning effect** (i.e. a moment).

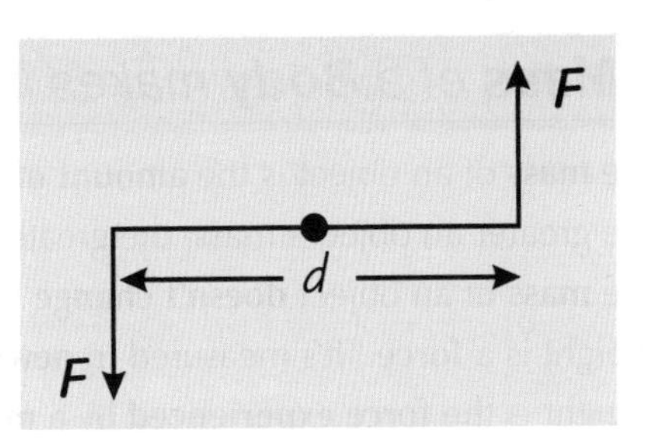

The **size** of this **moment** depends on the **size** of the **forces** and the **distance** between them.

moment of a couple (in Nm) = **size of one of the forces** (in N) × **perpendicular distance between the lines of action of the forces** (in m)

Again, in symbols, that's: $M = F \times d$

Example:

A cyclist turns a sharp right corner by applying equal but opposite forces of 20 N to the ends of the handlebars.

The length of the handlebars is 0.6 m.
Calculate the moment applied to the handlebars.

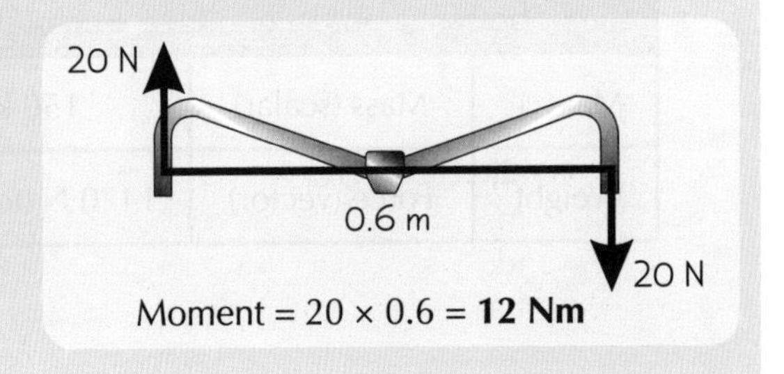

Moment = 20 × 0.6 = **12 Nm**

Practice Questions

Q1 A force of 54 N acts at a perpendicular distance of 84 cm from a pivot. Calculate the moment of the force.

Q2 A girl of mass 40 kg sits 1.5 m from the middle of a seesaw.
Show that her brother, mass 50 kg, must sit 1.2 m from the middle if the seesaw is to balance.

Q3 What is meant by the word 'couple'?

Q4 A racing car driver uses both hands to apply equal and opposite forces of 65 N to the edge of a steering wheel with radius 20 cm. Calculate the moment of the forces.

Exam Questions

Q1 A driver is changing his flat tyre. The moment required to undo the nut is 60 Nm.
He uses a 0.40 m long double-ended wheel wrench.
Calculate the force that he must apply at each end of the wrench. [2 marks]

Q2 A diver of mass 60 kg stands on the end of a diving board 2.0 m from the pivot point.
Calculate the downward force exerted on the board by the retaining spring 30 cm from the pivot.

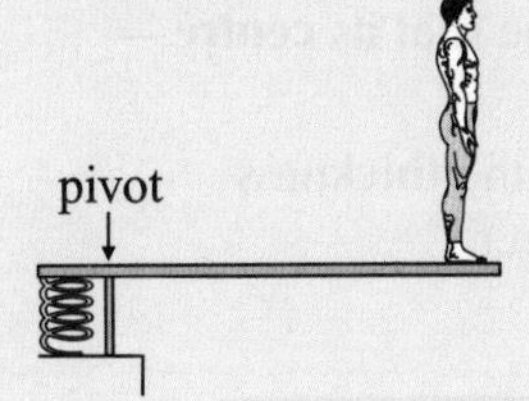

[2 marks]

It's all about balancing — just ask a tightrope walker...

They're always boring questions aren't they — seesaws or bicycles. It'd be nice if just once, they'd have a question on... I don't know, rotating knives or something. Just something unexpected. It'd make physics a lot more fun, I'm sure.

Mass, Weight and Centre of Mass

I'm sure you know all this 'mass' and 'weight' stuff from GCSE. But let's just make sure...

The Mass of a Body makes it Resist Changes in Motion

1) The **mass** of an object is the **amount of 'stuff'** (or **matter**) in it. It's measured in **kg**.
2) The greater an object's mass, the greater its **resistance** to a **change in velocity** (called its **inertia**).
3) The **mass** of an object **doesn't change** if the strength of the **gravitational field** changes.
4) Weight is a **force**. It's measured in **newtons** (N), like all forces.
5) Weight is the **force experienced by a mass** due to a **gravitational field**.
6) The weight of an object **does vary** according to the size of the **gravitational field** acting on it.

weight = mass × gravitational field strength ($W = mg$) where g = 9.81 Nkg^{-1} on Earth.

This table shows Derek (the lion)'s mass and weight on the Earth and the Moon.

Name	Quantity	Earth (g = 9.81 Nkg^{-1})	Moon (g = 1.6 Nkg^{-1})
Mass	Mass (scalar)	150 kg	150 kg
Weight	Force (vector)	1470 N (to 3 s.f.)	240 N (to 2 s.f.)

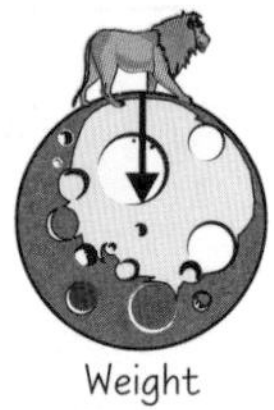

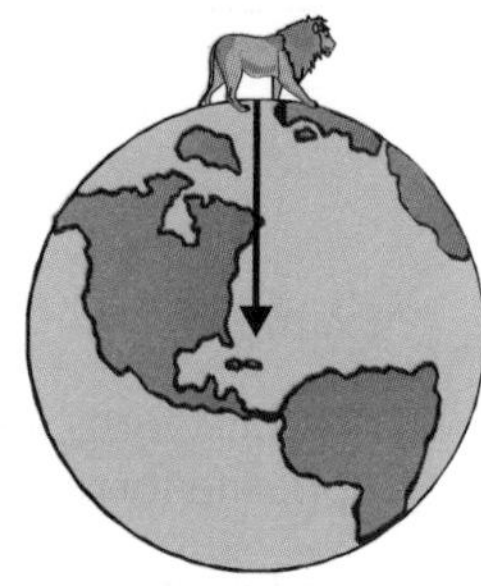

Centre of Mass — Assume All the Mass is in One Place

1) The **centre of mass** of an object is the **single point** that you can consider its **whole weight** to **act through** (whatever its orientation).
2) The object will always **balance** around this **point**, although in some cases the **centre of mass** will **fall outside** the object.
3) The centre of mass of a **uniform, regular solid** (e.g. a sphere, a cube) is at its **centre**.

Find the Centre of Mass Either by Symmetry...

1) To find the centre of mass for a **regular** object you can just use **symmetry**.
2) The centre of mass of any regular shape is at its **centre** — where the lines of symmetry will cross.
3) The centre of mass is **halfway** through the **thickness** of the object at the point the lines meet.

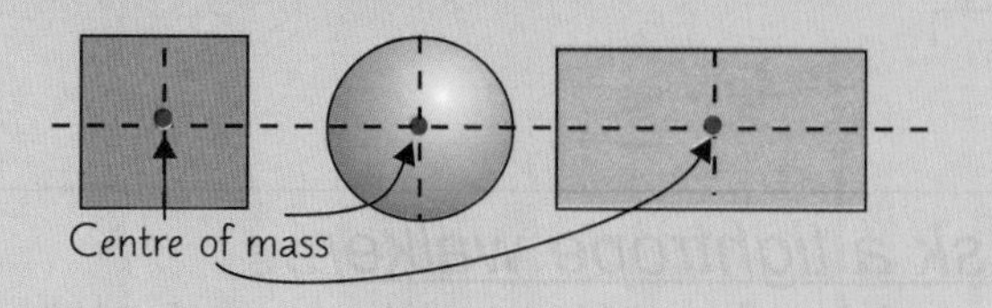

The symmetry in this picture shows the centre of cuteness.

Mass, Weight and Centre of Mass

... Or By *Experiment*

Experiment to find the Centre of Mass of an Irregular Object

1) **Hang** the object freely from a point (e.g. one corner).
2) Draw a **vertical line** downwards from the point of suspension — use a plumb bob to get your line exactly vertical.
3) Hang the object from a different point.
4) Draw another vertical line down.
5) The centre of mass is where the two lines **cross**.

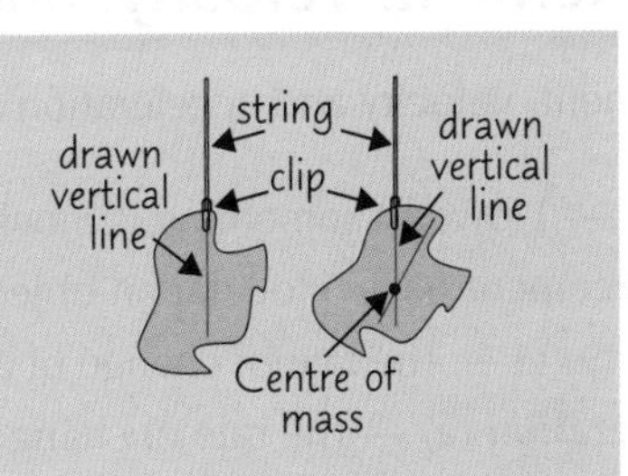

How *Stable* an Object is Depends on its *Centre of Mass* and Base Area

1) An object will topple over if a **vertical line** drawn **downwards** from its **centre of mass** (i.e. the line of action of its weight) falls **outside** its **base area**.
2) This is because a **resultant moment** (page 42) occurs, which provides a **turning force**.
3) An object will be nice and **stable** if it has a **low centre of mass** and a **wide base area**. This idea is used a lot in design, e.g. Formula 1® racing cars.
4) The higher the centre of mass and the smaller the base area, the less stable the object is. Think of unicyclists...

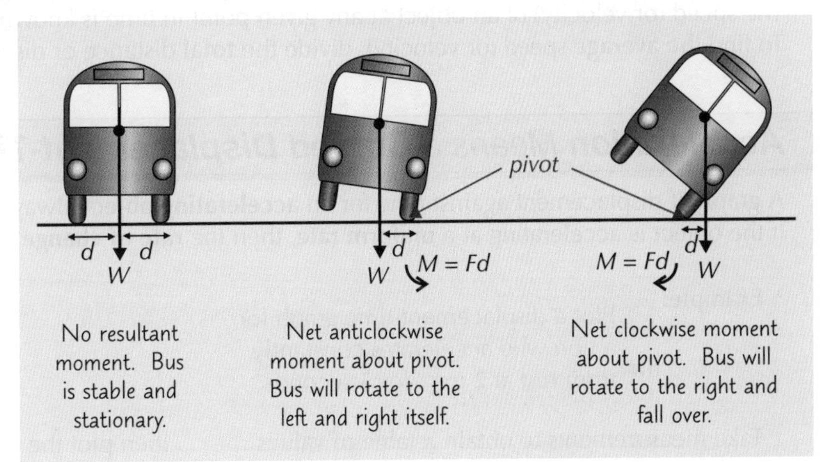

No resultant moment. Bus is stable and stationary.

Net anticlockwise moment about pivot. Bus will rotate to the left and right itself.

Net clockwise moment about pivot. Bus will rotate to the right and fall over.

Practice Questions

Q1 What are the differences between mass and weight?

Q2 A lioness has a mass of 200 kg. What would be her mass and weight on the Earth (where $g = 9.81$ Nkg^{-1}) and on the Moon (where $g = 1.6$ Nkg^{-1})?

Q3 What is meant by the centre of mass of an object?

Q4 Why will an object topple if its centre of mass is not over the object's base?

Exam Question

Q1 a) Describe an experiment to find the centre of mass of an object of uniform density with a constant thickness and irregular cross-section. Identify one major source of uncertainty and suggest a way to reduce its effect on your result. [5 marks]

b) Explain why you would not need to conduct this experiment for a regular, uniform solid. [1 mark]

The centre of mass of this book should be round about page 52...

This is a really useful area of physics. To would-be nuclear physicists it might seem a little dull, but if you want to be an engineer — something a bit more useful (no offence Einstein) — then things like centre of mass are dead important things to understand. You know, for designing things like cars and submarines... yep, pretty useful I'd say.

Displacement-Time Graphs

Drawing graphs by hand — oh joy. You'd think examiners had never heard of the graphical calculator.
Ah well, until they manage to drag themselves out of the dark ages, you'll just have to grit your teeth and get on with it.

Displacement, Velocity, *and* ***Acceleration*** *are All Linked*

Displacement, velocity and acceleration are all **vector** quantities (page 38), so the direction matters.

Speed — How fast something is moving, regardless of direction.
Displacement (s) — How far an object's travelled from its starting point in a given direction.
Velocity (v) — The rate of change of an object's displacement (its speed in a given direction).
Acceleration (a) — The rate of change of an object's velocity.

You need to know these formulas for **velocity** and **acceleration**:

$$v = \frac{\Delta s}{\Delta t} \quad \text{and} \quad a = \frac{\Delta v}{\Delta t}$$

The triangle symbol is the Greek 'delta', and it means 'the change in'.

The speed (or velocity) of an object at any given point in time is known as its **instantaneous** speed (or velocity).
To find the **average** speed (or velocity), divide the **total** distance or displacement by the total time.

Acceleration *Means a* ***Curved Displacement-Time Graph***

A graph of displacement against time for an **accelerating object** always produces a **curve**.
If the object is accelerating at a **uniform rate**, then the **rate of change** of the **gradient** will be constant.

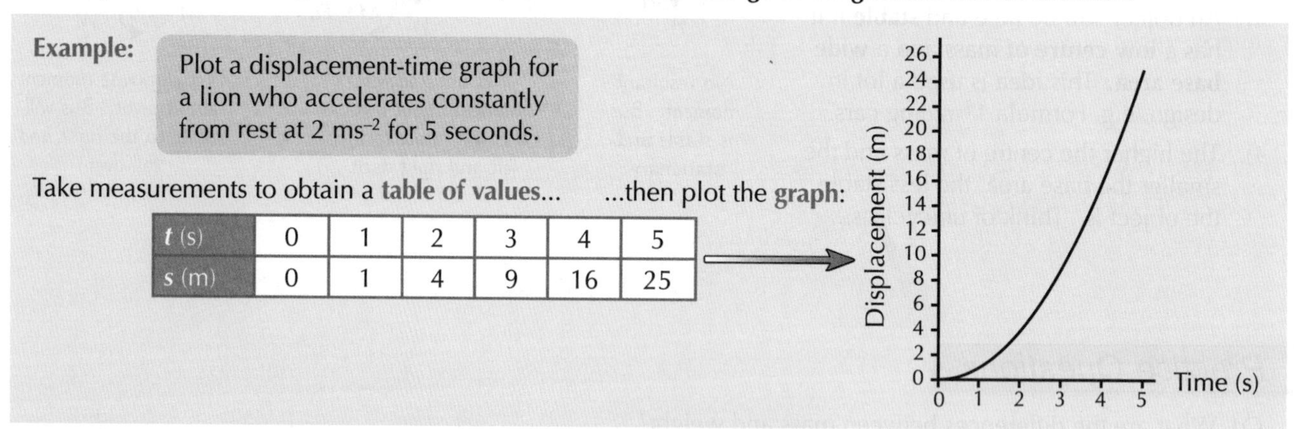

Example: Plot a displacement-time graph for a lion who accelerates constantly from rest at 2 ms^{-2} for 5 seconds.

Take measurements to obtain a **table of values**...

t (s)	0	1	2	3	4	5
s (m)	0	1	4	9	16	25

...then plot the **graph**:

Different Accelerations Have ***Different Gradients***

In the example above, if the lion has a **different acceleration** it'll change the **gradient** of the curve like this:

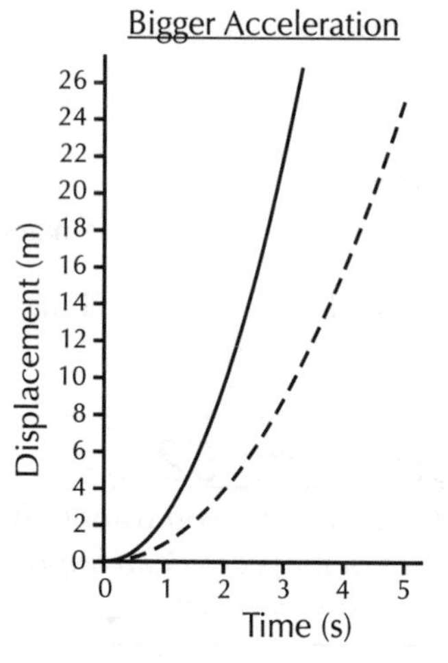

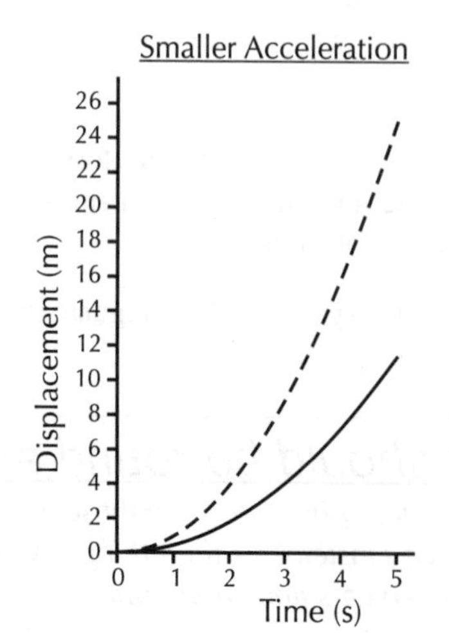

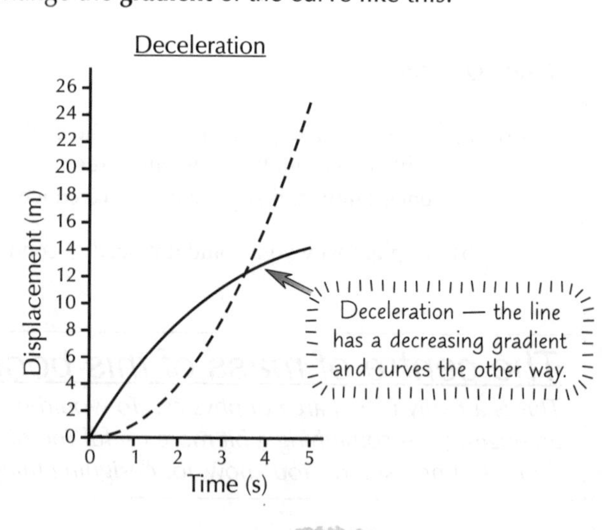

Displacement-Time Graphs

The Gradient of a Displacement-Time Graph Tells You the Velocity

When the velocity is constant, the graph's a **straight line**.
As you saw on the previous page, velocity is defined as...

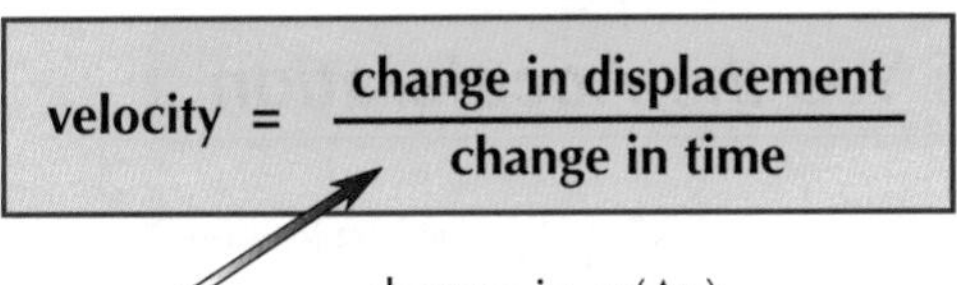

On the graph, this is $\frac{\text{change in } y\ (\Delta y)}{\text{change in } x\ (\Delta x)}$, i.e. the gradient.

So to get the velocity from a displacement-time graph, just find the gradient.

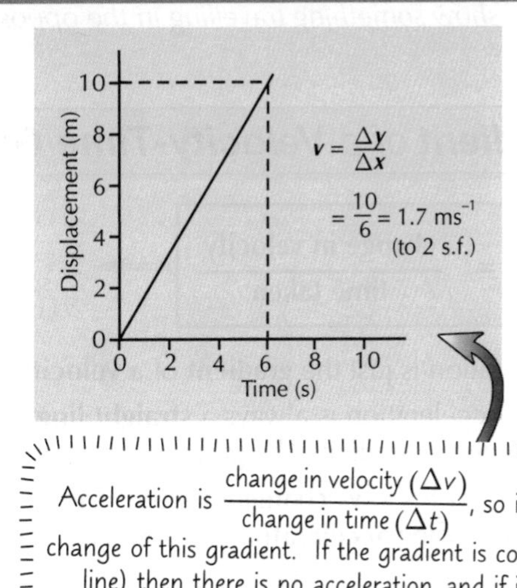

Acceleration is $\frac{\text{change in velocity } (\Delta v)}{\text{change in time } (\Delta t)}$, so it is the rate of change of this gradient. If the gradient is constant (straight line) then there is no acceleration, and if it's changing (curved line) then there's acceleration or deceleration.

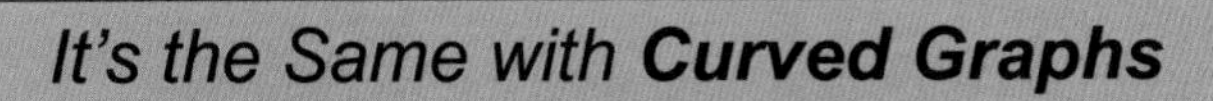

It's the Same with Curved Graphs

If the gradient **isn't constant** (i.e. if it's a curved line), it means the object is **accelerating**.

To find the **instantaneous velocity** at a certain point you need to draw a **tangent** to the curve at that point and find its gradient.

To find the **average velocity** over a period of time, just divide the total change in displacement by the total change in time — it doesn't matter if the graph is curved or not.

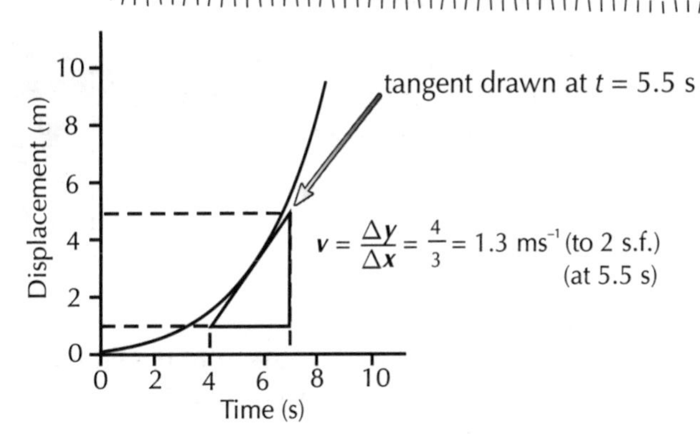

Practice Questions

Q1 What is given by the slope of a displacement-time graph?

Q2 Sketch a displacement-time graph to show: a) constant velocity, b) acceleration, c) deceleration

Exam Questions

Q1 Describe the motion of the cyclist as shown by the graph below. [4 marks]

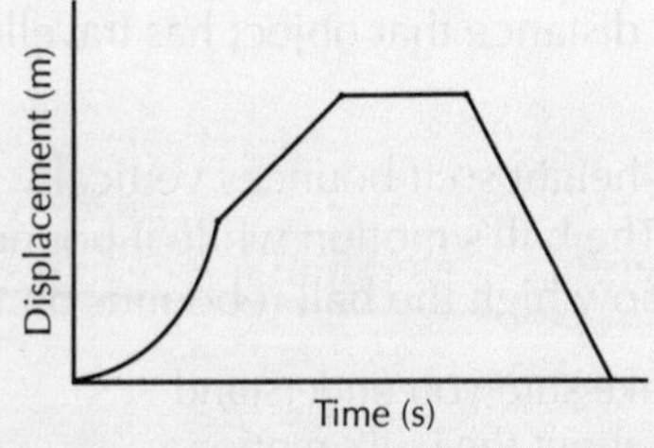

Q2 A baby crawls 5 m in 8 seconds at a constant velocity. She then rests for 5 seconds before crawling a further 3 m in 5 seconds. Finally, she makes her way back to her starting point in 10 seconds, travelling at a constant speed all the way.

a) Draw a displacement-time graph to show the baby's journey. [4 marks]

b) Calculate her average velocity at all the different stages of her journey. [2 marks]

Some curves are bigger than others...

Whether it's a straight line or a curve, the steeper it is, the greater the velocity. There's nothing difficult about these graphs — the problem is that it's easy to get them muddled up with velocity-time graphs (next page). Just think about the gradient — is it velocity or acceleration, is it changing (curve) or constant (straight line), is it 0 (horizontal line)...

Velocity-Time and Acceleration-Time Graphs

Speed-time graphs and velocity-time graphs are pretty similar. The big difference is that velocity-time graphs can have a negative part to show something travelling in the opposite direction:

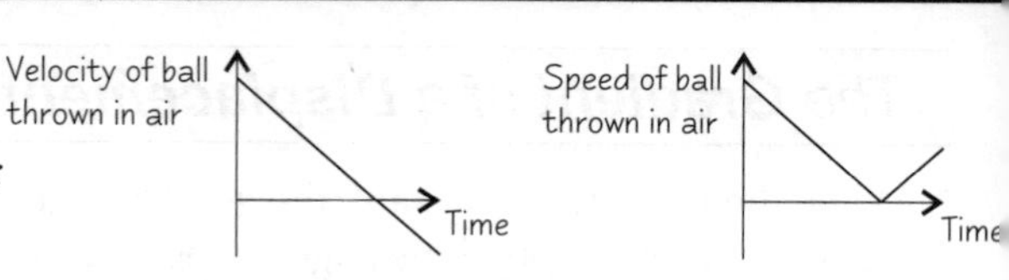

*The **Gradient** of a **Velocity-Time Graph** Tells You the **Acceleration***

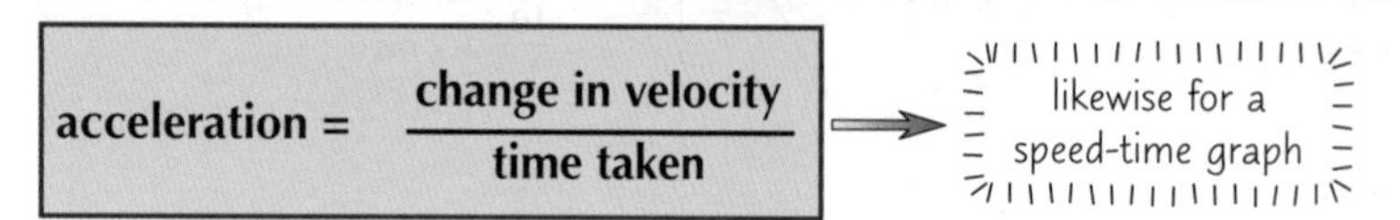

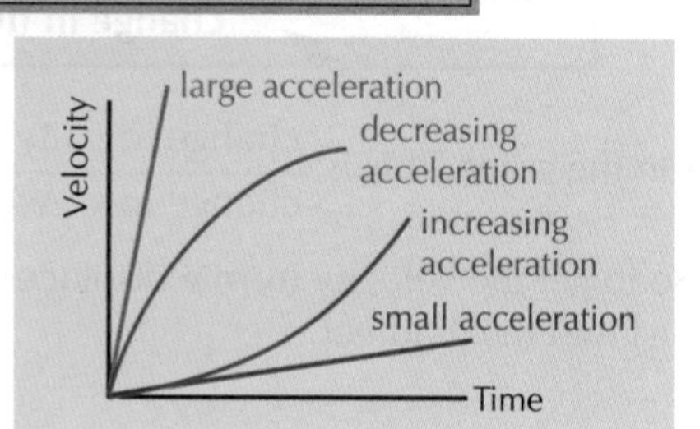

So the acceleration is just the **gradient** of a **velocity-time graph**.

1) **Uniform** acceleration is always a **straight line**. The **steeper** the **gradient**, the **greater** the **acceleration**.
2) A **curved** graph shows **changing** acceleration. **Increasing** gradient means **increasing acceleration**, and **decreasing** gradient means **decreasing acceleration** (or deceleration).

Example: A lion strolls along at 1.5 ms^{-1} for 4 s and then accelerates uniformly at a rate of 2.5 ms^{-2} for 4 s. Plot this information on a velocity-time graph.

So, for the first four seconds, the velocity is 1.5 ms^{-1}, then it increases by **2.5 ms^{-1} every second**:

t (s)	v (ms^{-1})
0 – 4	**1.5**
5	**4.0**
6	**6.5**
7	**9.0**
8	**11.5**

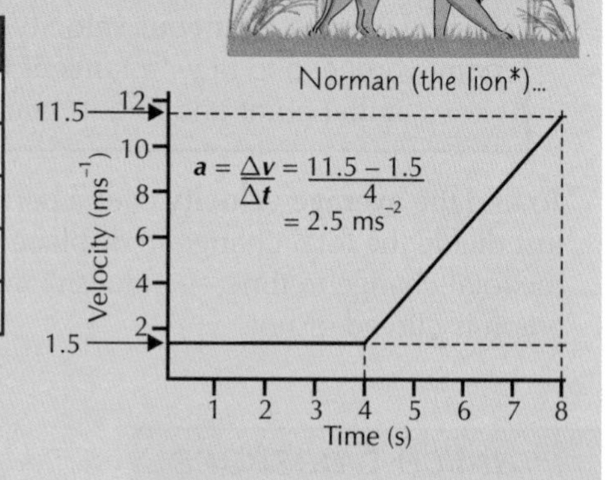

You can see that the **gradient of the line** is **constant** between 4 s and 8 s and has a value of 2.5 ms^{-2}, representing the **acceleration of the lion.**

*Yes, I know — I just like lions, OK...

Displacement** = **Area** Under **Velocity-Time Graph

You know that: **distance travelled = average speed × time**

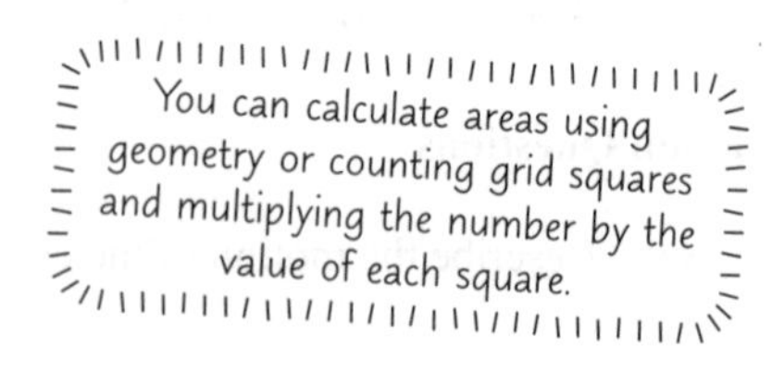

The **area under a velocity-time graph** tells you the displacement of the object. The magnitude of this displacement is the distance that object has travelled.

Example: A ball is dropped from table-height so it bounces vertically. It bounces twice before someone catches it. The ball's motion while it bounces is shown on the v-t graph below. Calculate how high the ball rebounds on the first bounce.

Before you try and calculate anything, make sure you understand what each part of the graph is telling you about the ball's motion.

1) When the ball is first dropped, the velocity of the ball is **negative** — so downwards is the negative direction.
2) The points where the **ball hits the floor** are shown by the **vertical straight lines** on the graph — the ball's **speed** remains roughly the **same**, but its **direction** (and **velocity**) changes the instant it hits the floor.
3) The points where the ball's velocity is **zero** show where the ball reaches the **top of a bounce** before starting to fall downwards.

The height of the first bounce is the **area under the graph** between the time the ball first rebounds from the floor and the time it reaches the top of the bounce.

displacement = area under graph
= (3.5 × 0.35) ÷ 2 = 0.6125 = **0.61 m (to 2 s.f.)**

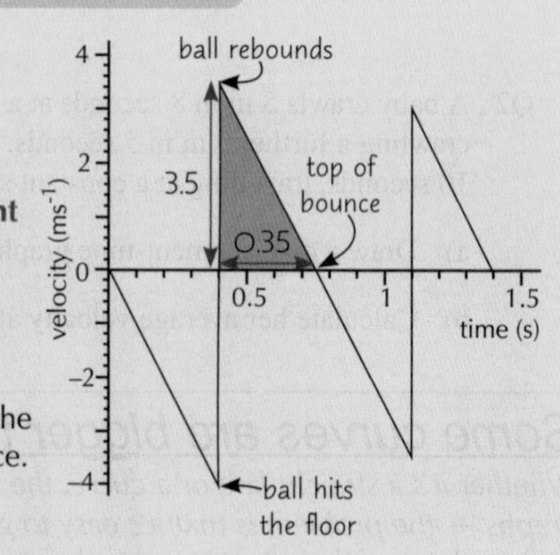

Velocity-Time and Acceleration-Time Graphs

Acceleration-Time Graphs are Useful Too

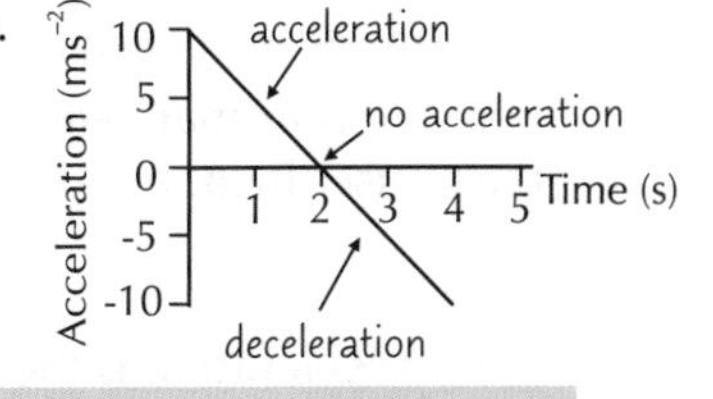

An **acceleration-time** (***a/t***) **graph** shows how an object's **acceleration** changes over time.

1) The **height** of the graph gives the object's **acceleration** at that time.
2) The **area** under the graph gives the object's **change in velocity**.
3) If **a = 0**, then the object is moving with **constant velocity**.
4) A negative acceleration is a **deceleration**.

Example: The acceleration of a car in a drag race is shown in this acceleration-time graph.

a) **After how many seconds does the car reach its maximum velocity?**
When the acceleration is 0, i.e. after **4 seconds**.

b) **If the car was stationary at t = 0 s, calculate its maximum velocity.**
(Change in) velocity = area under graph
= 0.5 × 4 × 25 = **50 ms^{-1}**

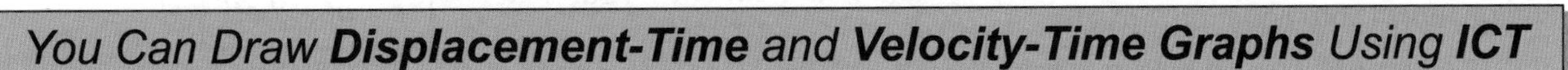

*You Can Draw **Displacement-Time** and **Velocity-Time Graphs** Using **ICT***

Instead of gathering distance and time data using **traditional methods**, e.g. a stopwatch and ruler, you can be a bit more **high-tech**.

A fairly **standard** piece of kit you can use for motion experiments is an **ultrasound position detector**. This is a type of **data-logger** that automatically records the **distance** of an object from the sensor several times a second.

If you attach one of these detectors to a computer with **graph-drawing software**, you can get **real-time** displacement-time and velocity-time graphs.

The main **advantages** of data-loggers over traditional methods are:

1) The data is more **accurate** — you don't have to allow for human reaction times.
2) Automatic systems have a much higher **sampling** rate than humans — most ultrasound position detectors can take a reading ten times every second.
3) You can see the data displayed in **real time**.

Practice Questions

Q1 How do you calculate acceleration from a velocity-time graph?

Q2 How do you calculate the displacement travelled from a velocity-time graph?

Q3 Sketch velocity-time graphs for constant velocity and constant acceleration.

Q4 Sketch velocity-time and acceleration-time graphs for a boy bouncing on a trampoline.

Q5 What does the area under an acceleration-time graph tell you?

Q6 Describe the main advantages of ICT over traditional methods for the collection and display of motion data.

Exam Question

Q1 A skier accelerates uniformly from rest at 2 ms^{-2} down a straight slope for 5 seconds. He then reaches the bottom of the slope and continues along the flat ground, decelerating at 1 ms^{-2} until he stops.

a) Sketch the velocity-time and acceleration-time graphs for his journey. [4 marks]

b) Use your v-t graph from part a) to find the distance travelled by the skier during the first 5 seconds. [2 marks]

Still awake — I'll give you five more minutes...

There's a really nice sunset outside my window. It's one of those ones that makes the whole landscape go pinky-yellowish. And that's about as much interest as I can muster on this topic. Normal service will be resumed on page 50.

Motion With Uniform Acceleration

Uniform Acceleration is Constant Acceleration

Acceleration could mean a change in speed or direction or both.

Uniform means **constant** here. It's nothing to do with what you wear.
There are **four main equations** that you use to solve problems involving **uniform acceleration**. You need to be able to **use them**, but you don't have to know how they're **derived** — we've just put it in to help you learn them.

1) **Acceleration is the rate of change of velocity.**
From this definition you get:

$a = \frac{(v - u)}{t}$ so $\boldsymbol{v = u + at}$

where:
u = initial velocity
v = final velocity
a = acceleration
t = time taken

2) **s = average velocity × time**
If acceleration is constant, the average velocity is just the average of the initial and final velocities, so:

$s = \frac{(u + v)}{2} \times t$ s = displacement

3) Substitute the expression for v from equation 1 into equation 2 to give:

$s = \frac{(u + u + at) \times t}{2} = \frac{2ut + at^2}{2}$ $\boldsymbol{s = ut + \frac{1}{2}at^2}$

4) You can **derive** the fourth equation from equations **1** and **2**:

Use equation **1** in the form: $a = \frac{(v - u)}{t}$

Multiply both sides by s, where: $s = \frac{(u + v)}{2} \times t$

This gives us: $as = \frac{(v - u)}{t} \times \frac{(u + v)t}{2}$

The t's on the right cancel, so:
$2as = (v - u)(v + u)$
$2as = v^2 - uv + uv - u^2$

so: $\boldsymbol{v^2 = u^2 + 2as}$

Example: A tile falls from a roof 25 m high. Calculate its speed when it hits the ground and how long it takes to fall. Take $g = 9.81\ \text{ms}^{-2}$.

First of all, write out what you know:
$s = 25$ m
$u = 0\ \text{ms}^{-1}$ since the tile's stationary to start with
$a = 9.81\ \text{ms}^{-2}$ due to gravity
$v = ?$ $t = ?$

Usually you take upwards as the positive direction. In this question it's probably easier to take downwards as positive, so you get $g = +9.81\ \text{ms}^{-2}$ instead of $g = -9.81\ \text{ms}^{-2}$.

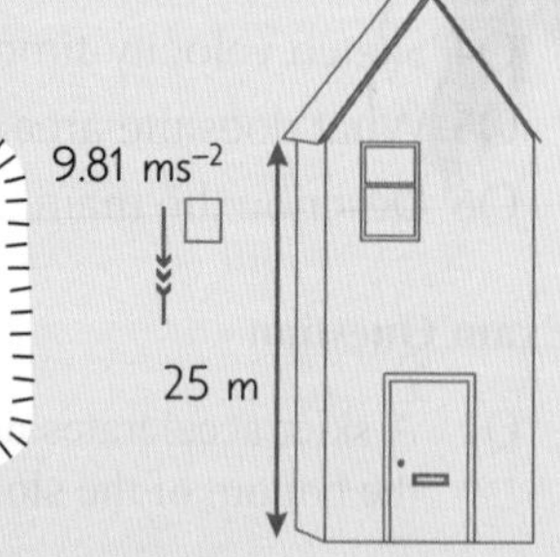

Then, choose an equation with only **one unknown quantity**.
So start with $v^2 = u^2 + 2as$
$v^2 = 0 + 2 \times 9.81 \times 25$
$v^2 = 490.5$
$v = 22\ \text{ms}^{-1}$ (to 2 s.f.)

Now, find t using:
$s = ut + \frac{1}{2}at^2$
$25 = 0 + \frac{1}{2} \times 9.81 \times t^2$
$t^2 = \frac{25}{4.9}$
$t = 2.3$ (to 2 s.f.)

Final answers:
t = **2.3 s (to 2 s.f.)**
v = **22 ms⁻¹ (to 2 s.f.)**

Motion With Uniform Acceleration

Example: A car accelerates steadily from rest at a rate of 4.2 ms^{-2} for 6.5 seconds.
a) Calculate the final speed.
b) Calculate the distance travelled in 6.5 seconds.

Remember — always start by writing down what you know.

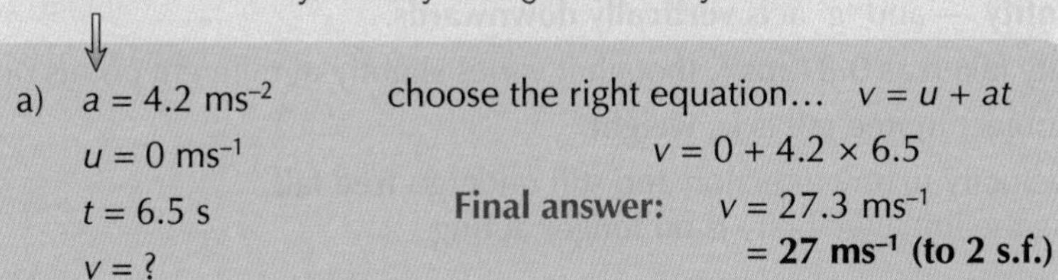

a) $a = 4.2$ ms^{-2}
$u = 0$ ms^{-1}
$t = 6.5$ s
$v = ?$

choose the right equation... $v = u + at$

$v = 0 + 4.2 \times 6.5$

Final answer: $v = 27.3$ ms^{-1}
$= \mathbf{27}$ **ms^{-1} (to 2 s.f.)**

b) $s = ?$
$t = 6.5$ s
$u = 0$ ms^{-1}
$a = 4.2$ ms^{-2}
$v = 27.3$ ms^{-1}

you can use: $s = \frac{(u+v)t}{2}$ ⇩ $s = \frac{(0 + 27.3) \times 6.5}{2}$ ⇩ **Final answer:** $s =$ **89 m (to 2 s.f.)**

or: $s = ut + \frac{1}{2}at^2$ ⇩ $s = 0 + \frac{1}{2} \times 4.2 \times (6.5)^2$ ⇩ $s =$ **89 m (to 2 s.f.)**

*You Have to **Learn** the Uniform Acceleration **Equations***

Make sure you learn the equations. There are only four of them and these questions are always dead easy marks in the exam, so you'd be dafter than a hedgehog in a helicopter not to learn them...

Practice Questions

Q1 Write out the four uniform acceleration equations.

Q2 A small steel ball is dropped from a height of 1.5 m. Calculate its speed as it hits the ground.

Exam Questions

Q1 A skydiver jumps from a helicopter hovering at a height of 1500 m from the ground.
She accelerates due to gravity for 5.0 s.
a) Calculate her maximum vertical velocity. (Assume no air resistance.) [2 marks]
b) Calculate how far she falls in this time. [2 marks]

Q2 A motorcyclist slows down uniformly as he approaches a red light.
He takes 3.2 seconds to come to a halt and travels 40 m (to 2 s.f.) in this time.
a) Calculate how fast he was initially travelling. [2 marks]
b) Calculate his acceleration. (N.B. a negative value shows a deceleration.) [2 marks]

Q3 A stream provides a constant acceleration of 6 ms^{-2}. A toy boat is pushed directly against the current and then released from a point 1.2 m upstream from a small waterfall. Just before it reaches the waterfall, it is travelling at a speed of 5 ms^{-1}.
a) Calculate the initial velocity of the boat. [2 marks]
b) Calculate the maximum distance upstream from the waterfall the boat reaches. [2 marks]

Constant acceleration — it'll end in tears...

If a question talks about "uniform" or "constant" acceleration, it's a dead giveaway they want you to use one of these equations. The tricky bit is working out which one to use — start every question by writing out what you know and what you need to know. That makes it much easier to see which equation you need. To be sure. Arrr.

Acceleration Due to Gravity

Ahhh acceleration due to gravity. The reason falling apples whack you on the head.

Free Fall is When There's Only *Gravity* and Nothing Else

Free fall is defined as the motion of an object undergoing an acceleration of 'g'. You need to remember:

1) Acceleration is a **vector quantity** — and 'g' acts **vertically downwards**.
2) The magnitude of 'g' is usually taken as **9.81 ms^{-2}**, though it varies slightly at different points on the Earth's surface.
3) The **only force** acting on an object in free fall is its **weight**.
4) Objects can have an initial velocity in any direction and still undergo **free fall** as long as the **force** providing the initial velocity is **no longer acting**.

All Objects in Free Fall *Accelerate* at the *Same* Rate

Another gravity experiment.

1) For over 1000 years the generally accepted theory was that heavier objects would fall towards the ground quicker than lighter objects. It was challenged a few times, but it was finally overturned when **Galileo** came on the scene.
2) The difference with Galileo was that he set up **systematic** and **rigorous experiments** to **test** his theories — just like in modern science. These experiments could be repeated and the results described **mathematically** and compared.
3) Galileo believed that all objects fall at the same rate. The problem in trying to prove it was that free-falling objects **fell too quickly** for him to be able to take any accurate measurements (he only had a water clock), and **air resistance** affects the rate at which objects fall.
4) Galileo measured the time a ball took to roll down a **smooth** groove in an inclined plane. He killed two birds with one stone by rolling it down a plane, which **slows** the ball's fall as well as reducing the effect of **air resistance**.
5) By rolling the ball along different fractions of the total length of the slope, he found that the distance the ball travelled was proportional to the square of the time taken. The ball was **accelerating** at a **constant rate**.
6) In the end it took **Newton** to bring it all together to show and explain why **all** free falling objects have the same acceleration. He showed **mathematically** that all objects are attracted towards the Earth due to a force he called **gravity**. Ah, good ol' Newton...

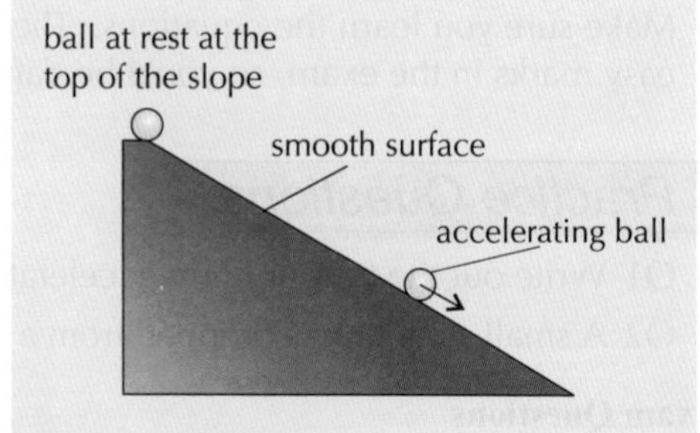

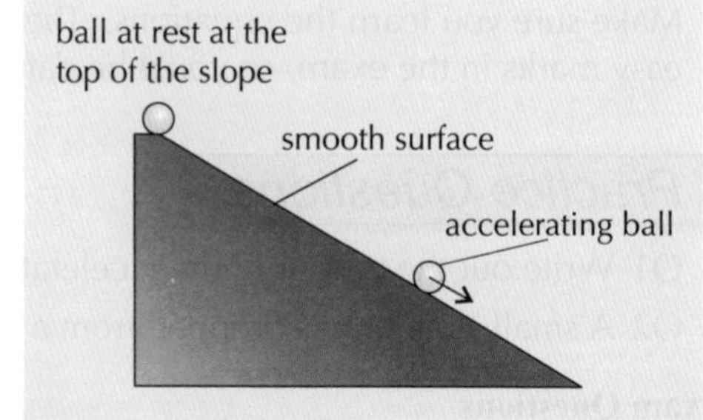

You Can Calculate **g** *By Performing an* ***Experiment...***

This is just one way of **measuring** g, there are loads of different experiments you could do — just make sure you know **one** method for your exams.

1) Set up the equipment shown in the diagram on the right.
2) Measure the height h from the **bottom** of the ball bearing to the **trapdoor**.
3) Flick the switch to simultaneously **start the timer** and **disconnect the electromagnet**, releasing the ball bearing.
4) The ball bearing falls, knocking the trapdoor down and **breaking the circuit** — which **stops the timer**. Record the time t shown on the timer.
5) **Repeat** this experiment three times and **average** the time taken to fall from this height. Repeat this experiment but drop the ball from several **different heights**.
6) You can then use these results to find g using a **graph** (see the next page).

electromagnet
ball bearing
switch
height h
timer
trapdoor

- Using a **small** and **heavy** ball bearing means you can assume air resistance is so small you can **ignore it**.
- Having a computer **automatically release** and **time** the ball-bearing's fall can measure times with a **smaller uncertainty** than if you tried to drop the ball and time the fall using a stopwatch.
- The most significant source of **error** in this experiment will be in the measurement of h. Using a ruler, you'll have an uncertainty of about ±1 mm. This dwarfs any error from switch delay or air resistance.

Acceleration Due to Gravity

...and Plotting a **Graph** of Your **Results**

1) Use your data from the experiment on the last page to plot a graph of **height** (s) against the **time** it takes the ball to fall, **squared** (t^2). Then draw a **line of best fit**.

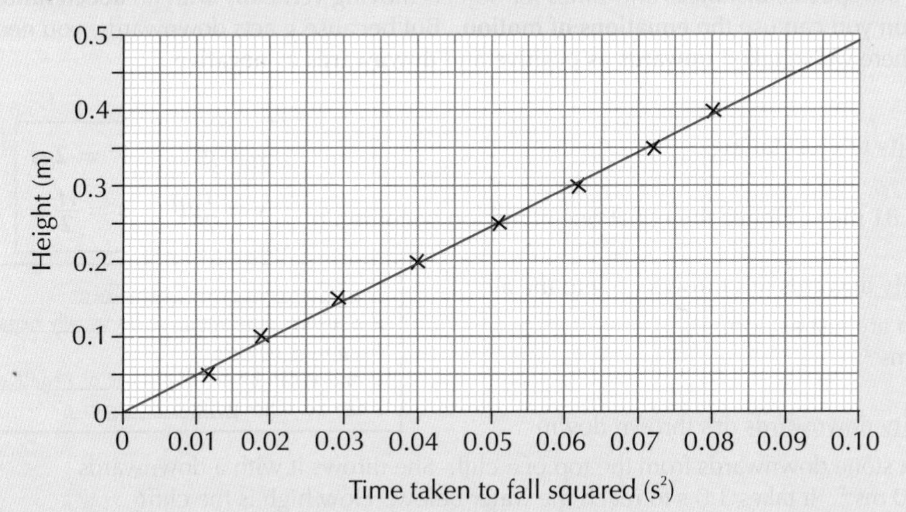

Before you do experiments, make sure you buff up on your practical skills — see pages 88-93.

You could plot error bars on your data graph to find the error in your final value for g. See page 91 for more on error bars.

2) You know that with constant acceleration, $s = ut + ½at^2$. If you drop the ball, initial speed $u = 0$, so $s = ½at^2$.
3) Rearranging this gives $½a = \frac{s}{t^2}$, or $½g = \frac{s}{t^2}$ (remember the acceleration is all due to gravity).
4) So the gradient of the line of best fit, $\frac{\Delta s}{\Delta t^2}$, is equal to $½g$:

$$g = 2 \times \frac{\Delta s}{\Delta t^2} = 2 \times \frac{0.44}{0.09} = \mathbf{9.8\ ms^{-2}} \text{ (to 2 s.f.)}$$

In the exam you might be asked to find g from a displacement-time graph (see p47). g is an acceleration and the gradient of the graph will be velocity, so you can find g by finding the change in gradient between two points on the graph (as $a = \Delta v \div \Delta t$).

Practice Questions

Q1 What is meant by free fall?

Q2 How does the velocity of a free-falling object change with time?

Q3 What is the main reason Galileo's ideas became generally accepted in place of the old theory?

Q4 Describe an experiment that could be used to calculate the value of g.

Exam Question

Q1 In an experiment to determine the value of g, a small steel ball is dropped from a range of heights. The time it takes to reach the ground when dropped from each height is recorded.

a) Explain why using a steel ball is better than using a beach ball in this experiment. [1 mark]

b) State one random error that could arise from this experiment and suggest a way to remove it. [2 marks]

c) State one systematic error that could arise from this experiment and suggest a way to remove it. [2 marks]

d) A graph of the distance travelled by the ball against time taken squared is plotted. Show that the gradient of the graph is equal to half the value of g. [3 marks]

So it's this "Galileo" geezer who's to blame for my practicals...

Hmmm... I wonder what Galileo would be proudest of — insisting on the systematic, rigorous experimental method on which modern science hangs... or getting in a Queen song? Magnificoooooo...

Projectile Motion

Calculators at the ready — it's time to resolve more things into vertical and horizontal components. It can be a bit tricky at first, but you'll soon get the hang of it. Chop chop, no time to lose.

You Can Just **Replace a** With **g** in the **Equations of Motion**

You need to be able to work out **speeds**, **distances** and **times** for objects moving vertically with an **acceleration** of g. As g is a **constant acceleration** you can use the **equations of motion**. But because g acts downwards, you need to be careful about directions, here we've taken **upwards as positive** and **downwards as negative**.

Case 1: No initial velocity (it's just falling)

Initial velocity $u = 0$

Acceleration $a = g = -9.81 \text{ ms}^{-2}$. Hence the equations of motion become:

$$v = gt \qquad v^2 = 2gs$$
$$s = \frac{1}{2}gt^2 \qquad s = \frac{vt}{2}$$

Case 2: An initial velocity upwards (it's thrown up into the air)

The equations of motion are just as normal, but with $a = g = -9.81 \text{ ms}^{-2}$.

Sign Conventions — Learn Them:
g is always downwards so it's usually negative
t is always positive
u and v can be either positive or negative
s can be either positive or negative

Case 3: An initial velocity downwards (it's thrown down)

Example: Alex throws a stone downwards from the top of a cliff. She throws it with a downwards velocity of 2.0 ms^{-1}. It takes 3.0 s to reach the water below. How high is the cliff?

1) You know $u = -2.0 \text{ ms}^{-1}$, $a = g = -9.81 \text{ ms}^{-2}$ and $t = 3.0$ s. You need to find s.

2) Use $s = ut + \frac{1}{2}gt^2 = (-2.0 \times 3.0) + (\frac{1}{2} \times -9.81 \times 3.0^2) = -50.145$ m. The cliff is **50 m (to 2 s.f.)** high.

s is negative because the stone ends up further down than it started. Height is a scalar quantity, so is always positive.

You Have to Think of **Horizontal** and **Vertical** Motion **Separately**

Example: Sharon fires a scale model of a TV talent show presenter horizontally with a velocity of 100 ms^{-1} (to 3 s.f.) from 1.5 m above the ground. How long does it take to hit the ground, and how far does it travel horizontally? Assume the model acts as a particle, the ground is horizontal and there is no air resistance.

Think about the vertical motion first:

1) It's **constant acceleration** under gravity...
2) You know $u = 0$ (no vertical velocity at first), $s = -1.5$ m and $a = g = -9.81 \text{ ms}^{-2}$. You need to find t.

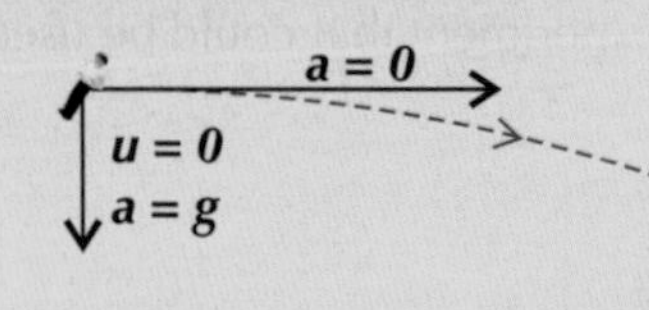

3) Use $s = \frac{1}{2}gt^2 \Rightarrow t = \sqrt{\frac{2s}{g}} = \sqrt{\frac{2 \times -1.5}{-9.81}} = 0.55300...$ s
4) So the model hits the ground after **0.55 seconds (to 2 s.f.)**.

Then do the horizontal motion:

1) The horizontal motion isn't affected by gravity or any other force, so it moves at a **constant speed**.
2) That means you can just use good old **speed = distance / time**.
3) Now $v_H = 100 \text{ ms}^{-1}$, $t = 0.55300...$ s and $a = 0$. You need to find s_H.
4) $s_H = v_H t = 100 \times 0.55300... = $ **55 m (to 2 s.f.)**

Where v_H is the horizontal velocity, and s_H is the horizontal distance travelled (rather than the height fallen).

Projectile Motion

It's Slightly Trickier if it Starts Off at an Angle

If something's projected at an angle (like, say, a javelin) you start off with both horizontal and vertical velocity:

Method:
1) Resolve the initial velocity into horizontal and vertical components.
2) Use the vertical component to work out how long it's in the air and/or how high it goes.
3) Use the horizontal component to work out how far it goes horizontally while it's in the air.

Example: A cannonball is fired from ground height at an angle of exactly 40° with an initial velocity of 15 ms^{-1}. Calculate how far the cannonball travels before it hits the ground. Assume no air resistance.

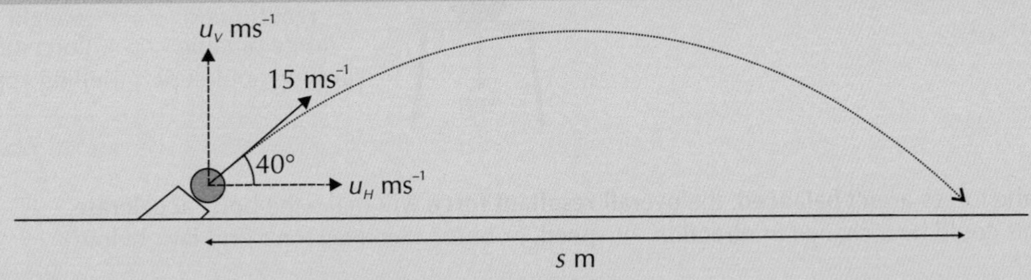

Resolve the velocity into horizontal and vertical components:

Horizontal component $u_H = \cos 40° \times 15 = 11.49...$ ms^{-1}

Vertical component $u_V = \sin 40° \times 15 = 9.64...$ ms^{-1}

Use the vertical component to work out how long the cannonball is in the air:

1) Halfway through the ball's flight, its v_V will be zero. $u_V = 9.64...$ ms^{-1}, $a = -9.81$ ms^{-2}, $t = ?$,

Use $v_V = u_V + at$: $0 = 9.64... + (-9.81 \times t) \Rightarrow t = \frac{9.64...}{9.81} = 0.98...$ s

2) So the time it takes to reach the ground again = $2 \times 0.98... = 1.96...$ s

Use the horizontal component to work out how far it goes while it's in the air:

There's no horizontal acceleration, so $u_H = V_H = 11.49...$ ms^{-1}.

Distance = constant speed × time = 11.49... × 1.96... = 22.58... = **23 m (to 2 s.f.)**

Practice Questions

Q1 What is the initial vertical velocity for an object projected horizontally with a velocity of 5 ms^{-1}?

Q2 What is the initial horizontal velocity of an object projected at 45 degrees to the ground with a velocity of 25 ms^{-1}?

Exam Questions

Q1 Jason stands on a vertical cliff edge throwing stones into the sea below. He throws a stone horizontally with a velocity of exactly 20 ms^{-1}, 560 m above sea level.

a) Calculate the time taken for the stone to hit the water from leaving Jason's hand. Use $g = 9.81$ ms^{-2} and ignore air resistance. [2 marks]

b) Calculate the distance of the stone from the base of the cliff when it hits the water. [2 marks]

Q2 Robin fires an arrow into the air with a vertical velocity of exactly 30 ms^{-1}, and a horizontal velocity of exactly 20 ms^{-1}, from 1 m above the ground. Calculate the maximum height from the ground reached by his arrow to the nearest metre. Use $g = 9.81$ ms^{-2} and ignore air resistance. [3 marks]

All this physics makes me want to create projectile motions...

...by throwing my revision books out of the window. The maths on this page can be tricky, but take it step by step and all will be fine. Plus, the next page is all about Newton, and I must say he was a mighty clever chap.

Newton's Laws of Motion

You did most of this at GCSE, but that doesn't mean you can just skip over it now. You'll be kicking yourself if you forget this stuff in the exam — it's easy marks...

Newton's **1st Law** Says That a **Force** is Needed to Change Velocity

1) **Newton's 1st law of motion** states that the **velocity** of an object will **not change** unless a **resultant force** acts on it.
2) In plain English this means a body will stay still or move in a **straight line** at a **constant speed**, unless there's a **resultant force** acting on it.

An apple sitting on a table won't go anywhere because the **forces** on it are **balanced**.

reaction (R) (force of table pushing apple up) = **weight** (mg) (force of gravity pulling apple down)

3) If the forces **aren't balanced**, the **overall resultant force** will make the body **accelerate**. This could be a change in **direction**, or **speed**, or both. (See Newton's 2nd law, below.)

Newton's **2nd Law** Says That **Acceleration** is **Proportional** to the Force

...which can be written as the well-known equation:

resultant force (N) = mass (kg) × acceleration (ms^{-2})

$$F = m \times a$$

Learn this — it crops up all over the place in Physics. And learn what it means too:

1) It says that the **more force** you have acting on a certain mass, the **more acceleration** you get.
2) It says that for a given force the **more mass** you have, the **less acceleration** you get.
3) There's more on this most excellent law on p.61.

REMEMBER:
1) The **resultant force** is the **vector sum** of all the forces.
2) The force is **always** measured in **newtons**.
3) The **mass** is always measured in **kilograms**.
4) The **acceleration** is always in the **same direction** as the **resultant force** and is measured in **ms^{-2}**.

Galileo said: **All Objects Fall** at the **Same Rate** (if You **Ignore Air Resistance**)

You need to understand **why** this is true. Newton's 2nd law explains it neatly — consider two balls dropped at the same time — ball **1** being heavy, and ball **2** being light. Then use Newton's 2nd law to find their acceleration.

mass = m_1
resultant force = F_1
acceleration = a_1
By Newton's Second Law:

$$F_1 = m_1a_1$$

Ignoring air resistance, the only force acting on the ball is weight, given by $W_1 = m_1g$ (where g = gravitational field strength = 9.81 Nkg^{-1}).
So: $F_1 = m_1a_1 = W_1 = m_1g$
So: $m_1a_1 = m_1g$, then m_1 cancels out to give: $a_1 = g$

W_1

mass = m_2
resultant force = F_2
acceleration = a_2
By Newton's Second Law:

$$F_2 = m_2a_2$$

Ignoring air resistance, the only force acting on the ball is weight, given by $W_2 = m_2g$ (where g = gravitational field strength = 9.81 Nkg^{-1}).
So: $F_2 = m_2a_2 = W_2 = m_2g$
So: $m_2a_2 = m_2g$, then m_2 cancels out to give: $a_2 = g$

W_2

...in other words, the **acceleration** is **independent of the mass**. It makes **no difference** whether the ball is **heavy or light**. And I've kindly **hammered home the point** by showing you two almost identical examples.

Newton's Laws of Motion

Newton's **3rd Law** Says Each Force has an **Equal, Opposite Reaction Force**

There are a few different ways of stating Newton's 3rd law, but the clearest way is:

> **If an object A EXERTS a FORCE on object B, then object B exerts AN EQUAL BUT OPPOSITE FORCE on object A.**

You'll also hear the law as "every action has an equal and opposite reaction". But this confuses people who wrongly think the forces are both applied to the same object. (If that were the case, you'd get a resultant force of zero and nothing would ever move anywhere...)

The two forces actually represent the **same interaction**, just seen from two **different perspectives**:

1) If you **push against a wall**, the wall will **push back** against you, **just as hard**. As soon as you stop pushing, so does the wall. Amazing...
2) If you **pull a cart**, whatever force **you exert** on the rope, the rope exerts the **exact opposite** pull on you (unless the rope's stretching).
3) When you go **swimming**, you push **back** against the water with your arms and legs, and the water pushes you **forwards** with an equal-sized force.

Newton's 3rd law applies in **all situations** and to all **types of force**. But the pairs of forces are always the **same type**, e.g. both gravitational or both electrical.

This looks like Newton's 3rd law...

But it's NOT.

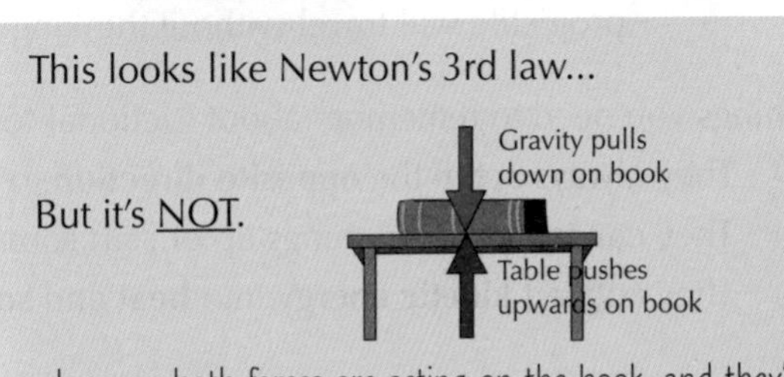

...because both forces are acting on the book, and they're not of the same type. They are two separate interactions. The forces are equal and opposite, resulting in zero acceleration, so this is showing Newton's 1st law.

Practice Questions

Q1 State Newton's 1st, 2nd and 3rd laws of motion, and explain what they mean.

Q2 What are the two equal and opposite forces acting between an orbiting satellite and the Earth?

Exam Questions

Q1 A boat is moving across a river. The engines provide a force of 500 N at right angles to the flow of the river and the boat experiences a drag of 100 N in the opposite direction. The force on the boat due to the flow of the river is 300 N. The mass of the boat is 250 kg.

a) Calculate the magnitude of the resultant force acting on the boat. [2 marks]

b) Calculate the magnitude of the acceleration of the boat. [2 marks]

Q2 John's bike, which has a mass of m, breaks and he has to push it home. The bike has a constant acceleration a and a frictional force F opposes the motion. What force is John is using to push his bike?

A	ma
B	$ma + F$
C	$m(a - F)$
D	$ma - F$

[1 mark]

Q3 Michael and Tom are both keen on diving. They notice that they seem to take the same time to drop from the diving board to the water. Use Newton's second law to explain why this is the case. (Assume no air resistance.) [3 marks]

Newton's three incredibly important laws of motion...

These laws may not really fill you with a huge amount of excitement (and I could hardly blame you if they don't)... but it was pretty fantastic at the time — suddenly people actually understood how forces work, and how they affect motion. I mean arguably it was one of the most important scientific discoveries ever...

Drag, Lift and Terminal Speed

If you jump out of a plane at 1500 m, you want to know that you're not going to be accelerating all the way.

Friction is a Force that Opposes Motion

There are two main types of friction — **dry friction** between **solid surfaces** and **fluid friction** (known as **drag**, fluid resistance or air resistance).

Fluid Friction or Drag:

1) 'Fluid' is a word that means either a **liquid or a gas** — something that can **flow**.
2) The force depends on the thickness (or **viscosity**) of the fluid.
3) It **increases** as the **speed increases** (for simple situations it's directly proportional, but you don't need to worry about the mathematical relationship).
4) It also depends on the **shape** of the object moving through it — the larger the **area** pushing against the fluid, the greater the resistance force.
5) A **projectile** (see p. 54) is **slowed down** by air resistance. If you calculate how far a projectile will travel without thinking about air resistance, your answer will be **too large**.

Things you need to remember about frictional forces:

1) They **always** act in the **opposite direction** to the **motion** of the object.
2) They can **never** speed things up or start something moving.
3) They convert **kinetic energy** into **heat** and **sound**.

Lift is Perpendicular to Fluid Flow

1) '**Lift**' is an **upwards force** on an object moving through a fluid.
2) It happens when the shape of an object causes the fluid flowing over it to **change direction**.
3) The force acts **perpendicular** to the direction the fluid flows in.

Example: Cross-section of a plane wing moving through air

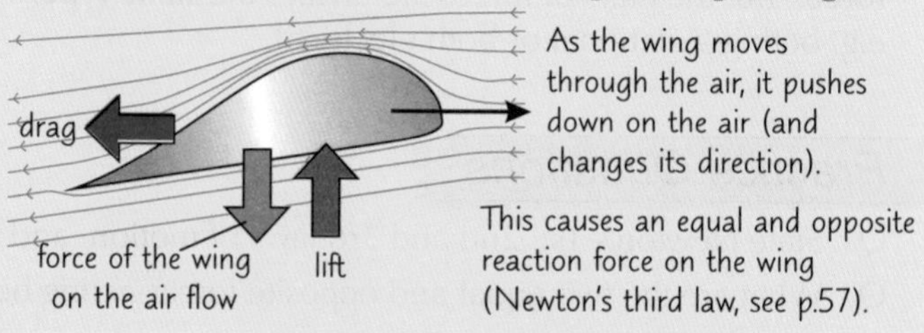

Terminal Speed — When the Friction Force Equals the Driving Force

You will reach a **terminal (maximum) speed** at some point, if you have:

1) a **driving force** that stays the **same** all the time
2) a **frictional** or **drag force** (or collection of forces) that increases with speed

There are **three main stages** to reaching terminal speed:

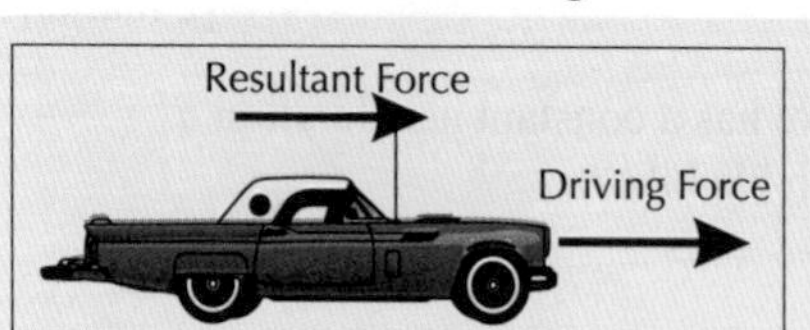

The car **accelerates** from **rest** using a constant driving force.

As the **speed increases**, the **frictional forces increase** (because of things like turbulence — you don't need the details). This **reduces the resultant force** on the car and hence **reduces its acceleration**.

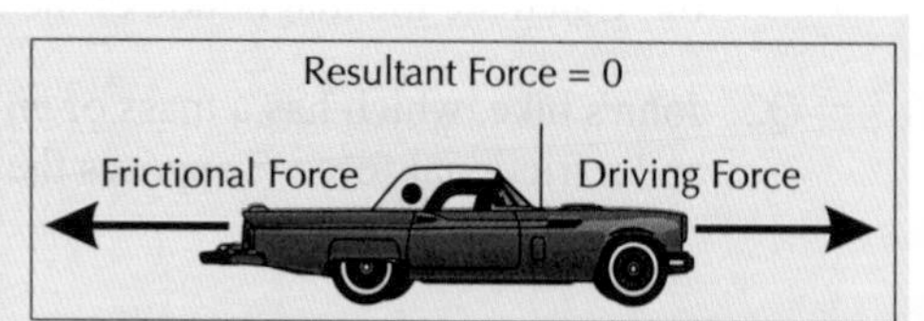

Eventually the car reaches a speed at which the **frictional forces are equal to the driving force**. There is now **no resultant force** and **no acceleration**, so the car carries on at **constant speed**.

Different factors affect a vehicle's maximum speed

As you just saw, a vehicle reaches maximum speed when the driving force is equal to the frictional force. So there are two main ways of increasing a vehicle's maximum speed:

1) **Increasing the driving force**, e.g. by increasing the engine size.
2) **Reducing the frictional force**, e.g. making the body more streamlined.

Drag, Lift and Terminal Speed

Things **Falling** through **Air** or **Water** Reach a **Terminal Speed** too

When something's falling through air, the weight of the object is a constant force accelerating the object downwards. Air resistance is a frictional force opposing this motion, which increases with speed.
So before a parachutist opens the parachute, exactly the same thing happens as with the car example:

1) A skydiver leaves a plane and will **accelerate** until the **air resistance** equals his **weight**.

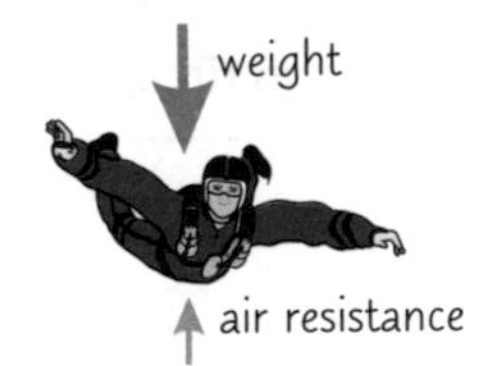

2) He will then be travelling at a **terminal speed**.

But... the terminal speed of a person in free fall is too great to land without dying a horrible death.
The **parachute increases** the **air resistance massively**, which slows him down to a lower terminal speed:

3) Before reaching the ground he will **open his parachute**, which immediately **increases the air resistance** so it is now **bigger** than his **weight**.

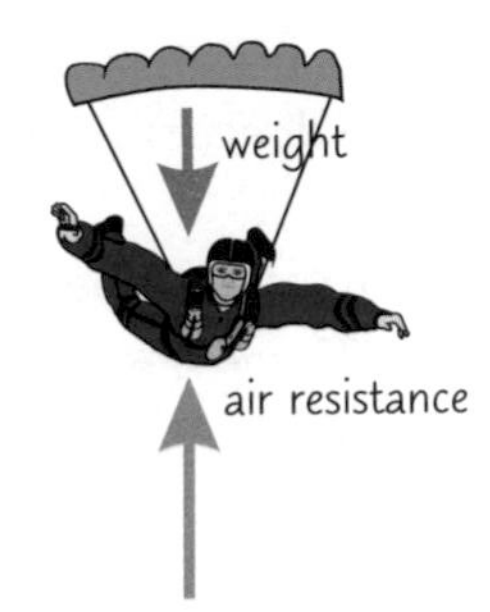

4) This **slows him down** until his speed has dropped enough for the **air resistance** to be **equal to his weight** again. This new terminal speed is small enough for him to land safely.

A *v-t* graph of the skydiver looks like this. He reaches terminal speed twice during his fall — the second one is much slower than the first.

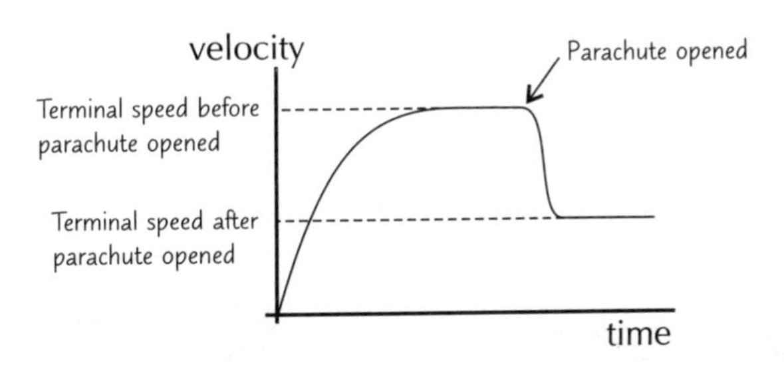

Practice Questions

Q1 What forces limit the speed of a skier going down a slope?

Q2 What causes a lift force on a plane wing as it moves through air?

Q3 Suggest two ways in which the maximum speed of a car can be increased.

Q4 What conditions cause a terminal speed to be reached?

Exam Question

Q1 A space probe free-falls towards the surface of a planet.
The graph on the right shows the velocity of the probe as it falls.

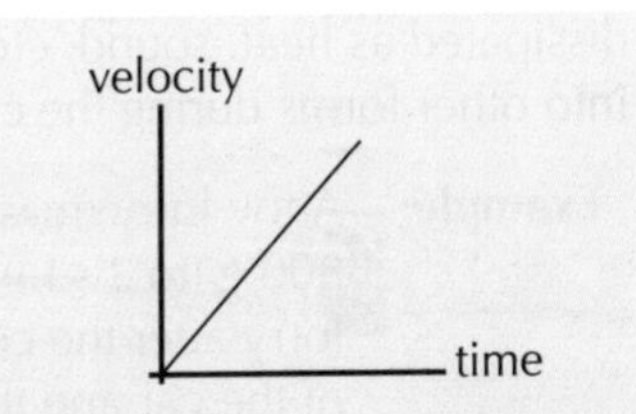

a) The planet does not have an atmosphere. Explain how you can tell this from the graph. [2 marks]

b) Sketch the velocity-time graph you would expect to see if the planet did have an atmosphere. [2 marks]

c) Explain the shape of the graph you have drawn. [3 marks]

*You'll never understand this without going parachuting...**

When you're doing questions about terminal velocity, remember the frictional forces reduce acceleration, not speed. They usually don't slow an object down, apart from in the parachute example, where the skydiver is travelling faster just before the parachute opens than the terminal velocity for the open parachute-skydiver combination.

* No. 37 in a series of the 100 least convincing excuses for an interesting holiday.

Momentum and Impulse

These pages are about linear momentum — that's momentum in a straight line (not a circle).

Understanding **Momentum** Helps You Do **Calculations** on **Collisions**

The **momentum** of an object depends on two things — its **mass** and **velocity**.
The **product** of these two values is the momentum of the object.

momentum = mass × velocity or in symbols: p **(in kg ms^{-1}) =** m **(in kg) ×** v **(in ms^{-1})**

Remember, momentum is a vector quantity, so it has size **and** direction.

Momentum is Always **Conserved**

You might see momentum referred to as 'linear momentum'. The other kind is 'angular momentum', but you don't need to know about that for now.

1) Assuming **no external forces** act, momentum is always **conserved**.
2) This means the **total momentum** of two objects **before** they collide **equals** the total momentum **after** the collision.
3) This is really handy for working out the **velocity** of objects after a collision (as you do...):

Example: A skater of mass 75 kg and velocity 4.0 ms^{-1} collides with a stationary skater of mass 50 kg (to 2 s.f.). The two skaters join together and move off in the same direction. Calculate their velocity after impact.

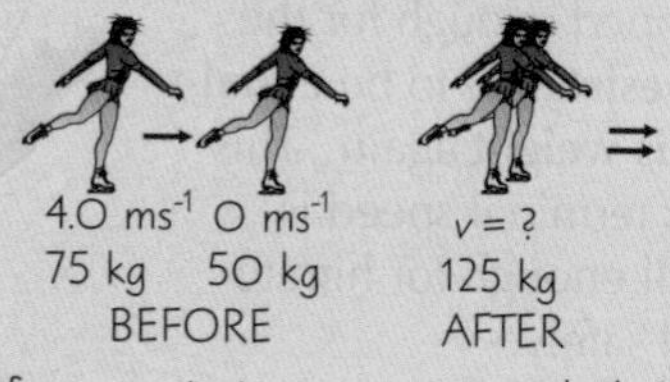

Before you start a momentum calculation, always draw a quick sketch.

Momentum of skaters before = Momentum of skaters after

$(75 \times 4.0) + (50 \times 0) = 125v$

$300 = 125v$

So v = **2.4 ms^{-1}**

4) The same principle can be applied in **explosions**. E.g. if you fire an **air rifle**, the **forward momentum** gained by the pellet **equals** the **backward momentum** of the rifle, and you feel the rifle recoiling into your shoulder.

Example: A bullet of mass 0.0050 kg is shot from a rifle at a speed of 200 ms^{-1}. The rifle has a mass of 4.0 kg. Calculate the velocity at which the rifle recoils.

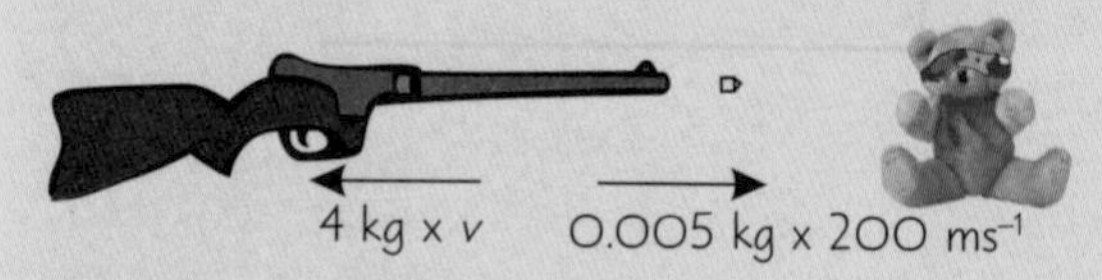

Momentum before explosion = Momentum after explosion

$0 = (0.0050 \times 200) + (4.0 \times v)$

$0 = 1 + 4v$

v = **−0.25 ms^{-1}**

Collisions Can be **Elastic** or **Inelastic**

An **elastic collision** is one where **momentum** is **conserved** and **kinetic energy** is **conserved** — i.e. no energy is dissipated as heat, sound, etc. If a collision is **inelastic** it means that some of the kinetic energy is converted into other forms during the collision. But **momentum is always conserved**.

Example: A toy lorry (mass 2.0 kg) travelling at 3.0 ms^{-1} crashes into a smaller toy car (mass 800 g (to 2 s.f.)), travelling in the same direction at 2.0 ms^{-1}. The velocity of the lorry after the collision is 2.6 ms^{-1} in the same direction. Calculate the new velocity of the car and the total kinetic energy before and after the collision.

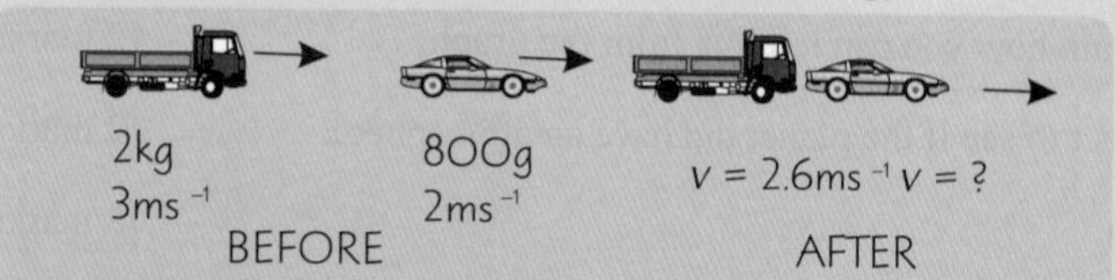

Momentum before collision = Momentum after collision

$(2 \times 3) + (0.8 \times 2) = (2 \times 2.6) + (0.8v)$

$7.6 = 5.2 + 0.8v$

$2.4 = 0.8v$

v = **3 ms^{-1}**

Kinetic energy before = KE of lorry + KE of car
$= \frac{1}{2}mv^2$ (lorry) $+ \frac{1}{2}mv^2$ (car)
$= \frac{1}{2}(2 \times 3^2) + \frac{1}{2}(0.8 \times 2^2)$
$= 9 + 1.6$
= **11 J (to 2 s.f.)**

Kinetic energy after $= \frac{1}{2}(2 \times 2.6^2) + \frac{1}{2}(0.8 \times 3^2)$
$= 6.76 + 3.6$
= **10 J (to 2 s.f.)**

The difference in the two values is the amount of kinetic energy dissipated as heat or sound, or in damaging the vehicles — so this is an **inelastic** collision.

Momentum and Impulse

Newton's 2nd Law Says That Force is the Rate of Change in Momentum...

The **rate of change of momentum** of an object is **directly proportional** to the **resultant force** which acts on the object.

So: $F = \frac{\Delta(mv)}{\Delta t}$ or $F\Delta t = \Delta(mv)$ (where F is constant)

Remember that acceleration is equal to the rate of change of velocity (page 48), so if mass is constant then this formula gives you that mechanics favourite, $F = m \times a$.

Impulse = Change in Momentum

1) Newton's second law says **force = rate of change of momentum**, or $F = (mv - mu) \div t$
2) **Rearranging** Newton's 2nd law gives: ⟹ $Ft = mv - mu$ so **impulse = change in momentum**
   Where **impulse** is defined as **force × time**, Ft.
   The units of impulse are **newton seconds**, Ns.

   Where v is the final velocity and u is the initial velocity.
3) **Impulse** is the **area under** a **force-time graph** — this is really handy for solving problems where the force changes.

Example: The graph shows the resultant force acting on a toy car. If the car is initially at rest, what is its momentum after 3 seconds?

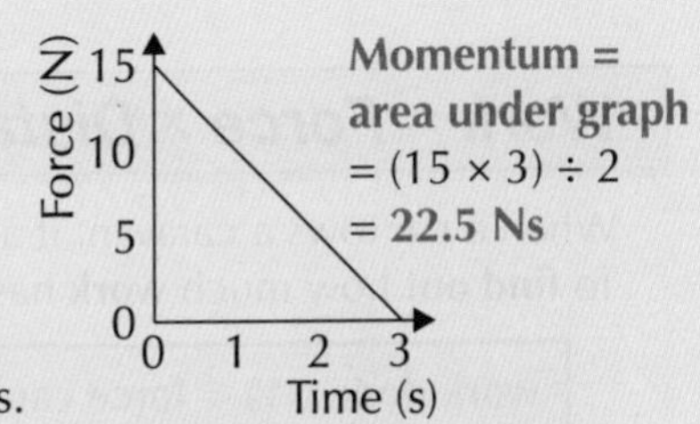

Momentum = area under graph
$= (15 \times 3) \div 2$
= 22.5 Ns

Impulse = change in momentum = $mv - mu$. The **initial momentum** (mu) is **zero** because the toy car is stationary to begin with. So, **impulse** = mv.

Impulse is the **area under the graph**, so to find the **momentum** of the car after 3 seconds, you need to find the **area under the graph** between 0 and 3 seconds.

4) The force of an impact is **increased** by **reducing** the impact time, e.g. the **less time** your foot is in contact with a football when kicking it, the **more force** you will kick it with (assuming the change in momentum is the same).
5) The **force** of an impact can be **reduced** by **increasing the time** of the impact, e.g. vehicle safety features.
6) In order to design vehicles **ethically**, manufacturers need to make sure the vehicles they produce are designed and fitted with features that help **protect people** in a crash.

- **Crumple zones** — the parts at the front and back of the car crumple up on impact. This causes the car to take longer to stop, increasing the impact time and decreasing the force on the passengers.
- **Seat belts** — these stretch slightly, increasing the time taken for the wearer to stop. This reduces the forces acting on the chest.
- **Air bags** — these also slow down passengers more gradually, and prevent them from hitting hard surfaces inside the car.

Practice Questions

Q1 Give two examples of conservation of momentum in practice.

Q2 Describe what happens when a tiny object makes an elastic collision with a massive object, and why.

Q3 Describe how seat belts reduce the force acting on a car passenger in a collision.

Exam Questions

Q1 A ball of mass 0.60 kg moving at 5.0 ms^{-1} collides with a larger stationary ball of mass 2.0 kg. The smaller ball rebounds in the opposite direction at 2.4 ms^{-1}.

a) Calculate the velocity of the larger ball immediately after the collision. [3 marks]

b) State and explain whether this is an elastic or inelastic collision. Support your answer with calculations. [3 marks]

Q2 A toy train of mass 0.7 kg, travelling at 0.3 ms^{-1}, collides with a stationary toy carriage of mass 0.4 kg. The two toys couple together. Calculate their new velocity. [3 marks]

Momentum'll never be an endangered species — it's always conserved...

*It seems odd to say that momentum's always conserved then tell you that impulse is the change in momentum. Impulse is just the change of momentum of one object, whereas conservation of momentum is talking about the **whole** system.*

Work and Power

As everyone knows, work in Physics isn't like normal work. It's harder. Work also has a specific meaning that's to do with movement and forces. You'll have seen this at GCSE — it just comes up in more detail for A level.

Work is Done Whenever Energy is Transferred

This table gives you some examples of **work being done** and the **energy changes** that happen.

1) Usually you need a force to move something because you're having to **overcome another force**.
2) The thing being moved has **kinetic energy** while it's **moving**.
3) The kinetic energy is transferred to **another form of energy** when the movement stops.

ACTIVITY	WORK DONE AGAINST	FINAL ENERGY FORM
Lifting up a box.	gravity	gravitational potential energy
Pushing a chair across a level floor.	friction	heat
Pushing two magnetic north poles together.	magnetic force	magnetic energy
Stretching a spring.	stiffness of spring	elastic potential energy

The word **'work'** in Physics means the **amount of energy transferred** from one form to another when a force causes a movement of some sort.

Work = Force × Distance

When a car tows a caravan, it applies a force to the caravan to move it to where it's wanted. To **find out** how much **work** has been **done**, you need to use the **equation**:

work done (W) = **force causing motion** (F) **× distance moved** (s), **or** $W = Fs$
...where W is measured in joules (J), F is measured in newtons (N) and s is measured in metres (m).

Points to remember:

1) **Work** is the **energy** that's been **changed** from one form to another — it's not necessarily the **total** energy. E.g. moving a book from a low shelf to a higher one will increase its gravitational potential energy, but it had some potential energy to start with. Here, the **work done** would be the **increase** in potential energy, **not the total** potential energy.
2) Remember the distance needs to be measured in metres — if you have **distance in centimetres or kilometres**, you need to **convert** it to metres first.
3) The force F will be a **fixed** value in any calculations, either because it's **constant** or because it's the **average** force.
4) The equation assumes that the **direction of the force** is the **same** as the **direction of movement**.
5) The equation gives you the **definition** of the joule (symbol J): 'One joule is the work done when a force of 1 newton moves an object through a distance of 1 metre'.

The Force isn't Always in the Same Direction as the Movement

Sometimes the **direction of movement** is **different** from the **direction of the force**.

Example:

1) To **calculate the work done** in a situation like the one on the right, you need to consider the **horizontal** and **vertical components** of the **force**.
2) The only **movement** is in the **horizontal** direction. This means the **vertical force** is not causing any motion (and hence not doing any work) — it's just **balancing** out some of the **weight**, meaning there's a **smaller reaction force**.

3) The horizontal force is causing the motion — so to **calculate** the **work done**, this is the **only force** you need to consider. Which means we get:

$$W = Fs\cos\theta$$

F
θ
F cos θ
Direction of motion

Where θ is the **angle** between the **direction of the force** and the **direction of motion**. See page 40 for more on resolving forces.

Work and Power

The Area Under a Force-Displacement Graph Tells You the Work Done

1) For a **variable force**, you can't just use the formula $W = Fs$ — nightmare.
2) Luckily, plotting a **graph** of **force against distance moved** lets you calculate the work done by just finding the **area under the graph**.
3) You might need to **split it up** into sections that make shapes you can work out the area for, e.g. trapeziums.

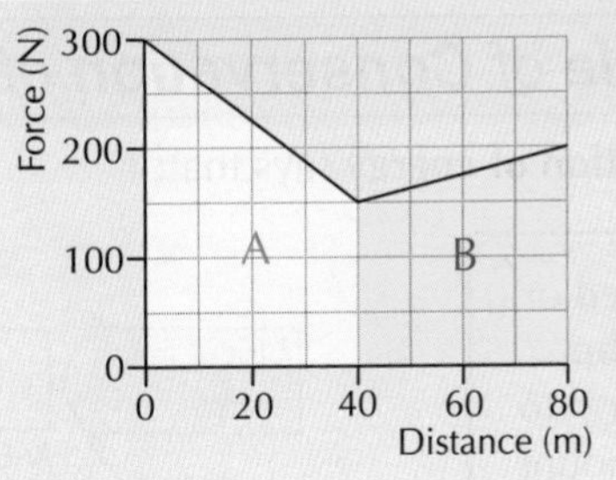

The graph shows the force exerted by Tibalt the circus monkey as he cycled up a hill.

Work done in section A:

$40 \times \frac{300 + 150}{2} = 9000$ J

Work done in section B:

$40 \times \frac{200 + 150}{2} = 7000$ J

Total work done = **16 000 J**

Power = Work Done per Second

Power means many things in everyday speech, but in Physics (of course!) it has a special meaning. Power is the **rate of doing work** — in other words it is the **amount of energy transferred** from one form to another **per second**. You **calculate power** from this equation:

> **Power (P) = change in energy (or work done) (ΔW) / change in time (Δt), or** $P = \frac{\Delta W}{\Delta t}$
>
> ...where P is measured in watts (W), ΔW is measured in joules (J) and Δt is measured in seconds (s).

The **watt** (symbol W) is defined as a **rate of energy transfer** equal to **1 joule per second** (Js^{-1}).

Power is also Force × Velocity

Sometimes, it's **easier** to use **this version** of the power equation:

1) You know $P = \Delta W / \Delta t$.
2) You also know $\Delta W = F\Delta s$, which gives $P = F\Delta s / \Delta t$.
3) But $v = \Delta s / \Delta t$, which you can substitute into the above equation to give:

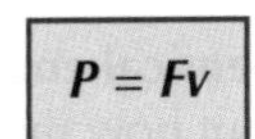

$$P = Fv$$

It's easier to use this if you're given speed in the question.

Practice Questions

Q1 Write down the equation used to calculate work if the force and motion are in the same direction.

Q2 Write down the equation for work if the force is at an angle to the direction of motion.

Q3 What does the area under a force-distance graph represent?

Q4 Write down the equations relating a) power and work and b) power and speed.

Exam Questions

Q1 A traditional narrowboat is drawn by a horse walking along the towpath. The horse pulls the boat at a constant speed between two locks which are 1500 m apart. The tension in the rope is exactly 100 N (to 2 s.f.) at exactly 40° to the direction of motion.

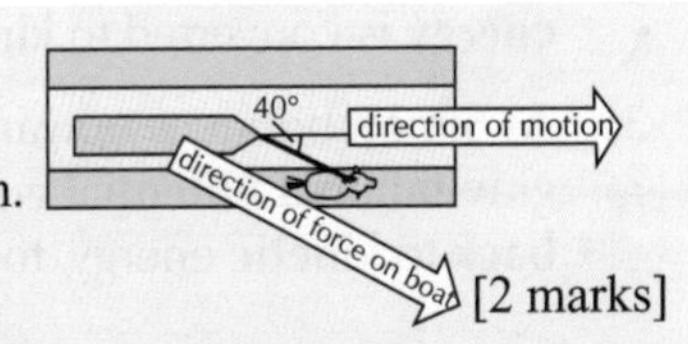

a) Calculate the work done on the boat. [2 marks]

b) The boat moves at $0.8\ ms^{-1}$. Calculate the power supplied to the boat in the direction of motion. [2 marks]

Q2 A motor is used to lift a 20 kg (to 2 s.f.) load a height of 3.0 m. (Take $g = 9.81\ Nkg^{-1}$.)

a) Calculate the work done in lifting the load. [2 marks]

b) The speed of the load during the lift is $0.25\ ms^{-1}$. Calculate the power delivered by the motor. [2 marks]

Work — there's just no getting away from it...

Loads of equations to learn. Well, that's what you came here for, after all. Can't beat a good bit of equation-learning, as I've heard you say quietly to yourself when you think no one's listening. Aha, can't fool me. Ahahahahahahahahahaha.

Conservation of Energy and Efficiency

Energy can never be lost. I repeat — energy can never be lost. Which is basically what I'm about to take up two whole pages saying. But that's, of course, because you need to do exam questions on this as well as understand the principle.

Learn the *Principle* of *Conservation* of *Energy*

The **principle of conservation of energy** says that:

Energy **cannot be created** or **destroyed**. Energy **can be transferred** from one form to another but the total amount of energy in a closed system will not change.

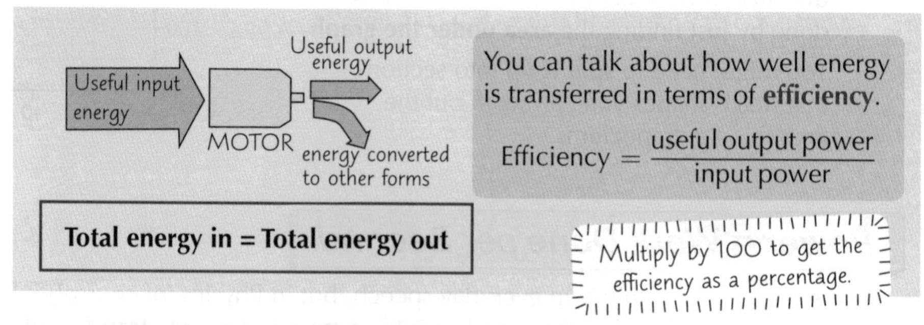

You can talk about how well energy is transferred in terms of **efficiency**.

$$\text{Efficiency} = \frac{\text{useful output power}}{\text{input power}}$$

Multiply by 100 to get the efficiency as a percentage.

You Need it for *Questions* about *Kinetic* and *Potential Energy*

The principle of conservation of energy nearly always comes up when you're doing questions about changes between kinetic and potential energy.

A quick reminder:

1) **Kinetic energy** is energy of anything **moving**, which you work out from $E_k = \frac{1}{2}mv^2$, where v is the velocity it's travelling at and m is its mass.
2) There are **different types of potential energy** — e.g. gravitational and elastic (see p.68).
3) **Gravitational potential energy** is the energy something gains if you lift it up. You work it out using: $\Delta E_p = mg\Delta h$, where m is the mass of the object, Δh is the height it is lifted and g is the gravitational field strength (9.81 Nkg^{-1} on Earth).
4) **Elastic strain energy** (elastic potential energy) is the energy you get in, say, a stretched rubber band or spring. You work this out using $E = \frac{1}{2}k\Delta l^2$, where Δl is the extension of the spring and k is the stiffness constant (p.66).

The energy you need to do things comes from your **food** — **chemical energy** inside the food is **converted** to other forms, e.g. **kinetic energy**. Be careful if you're trying to work out how much kinetic energy you can get from food though — a lot of the energy in food will actually be converted to other forms, e.g. **heat energy** to keep warm.

Examples: These pictures show you three **examples** of changes between kinetic and potential energy.

1) As Becky throws the **ball upwards**, **kinetic energy** is converted into **gravitational potential energy**. When it **comes down** again, that **gravitational potential** energy is **converted back** into **kinetic** energy.
2) As Dominic goes **down the slide**, **gravitational potential energy** is converted to **kinetic energy**.

3) As Simon bounces upwards from the trampoline, **elastic potential energy** is converted to **kinetic energy**, to **gravitational potential energy**. As he comes back down again, that **gravitational potential** energy is **converted back** to **kinetic** energy, to **elastic potential** energy, and so on.

In **real life** there are also **frictional forces** — Simon would have to exert some **force** from his **muscles** to keep **jumping** to the **same height** above the trampoline each time. Each time Simon jumps, some kinetic energy is converted to heat energy due to air resistance. You're usually told to **ignore friction** in exam questions — this means you can **assume** that the **only forces** are those that provide the **potential or kinetic energy** (in this example that's **Simon's weight** and the **tension** in the springs and trampoline material). If you're ignoring friction, you can say that the **sum of the kinetic and potential energies is constant.**

4) In a **car crash**, a lot of kinetic energy is transferred in a **short space of time**. Car safety features are designed to transfer some of this energy into other forms — this reduces the amount of energy transferred to the car passengers and other road users to help protect them. For example, **crumple zones** (see page 67) absorb some of the car's kinetic energy by **deforming**, **seat belts** absorb some of the passengers' kinetic energy by **stretching**.

Conservation of Energy and Efficiency

Use Conservation of Energy to **Solve Problems**

You need to be able to **use** conservation of mechanical energy (change in potential energy = change in kinetic energy) to solve problems. The classic example is the **simple pendulum**.

In a simple pendulum, you assume that all the mass is in the **bob** at the end.

Example: A simple pendulum has a mass of 720 g and a length of 50 cm (to 2 s.f.). It is pulled out to an angle of 30° (to 2 s.f.) from the vertical.

a) Find the gravitational potential energy stored in the pendulum bob.

You can work out the increase in height, Δh, of the end of the pendulum using trig.

Gravitational potential energy $= mg\Delta h$

$= 0.72 \times 9.81 \times (0.5 - 0.5 \cos 30°)$

$= 0.473...$ J

$=$ **0.47 J (to 2.s.f)**

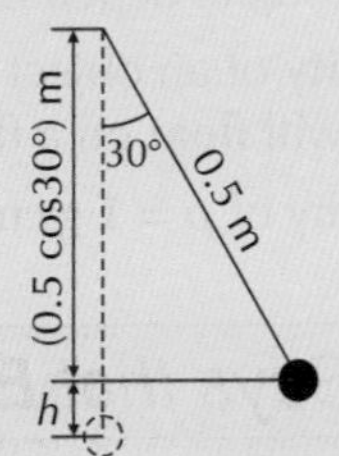

b) The pendulum is released. Find the maximum speed of the pendulum bob as it passes the vertical position.

To find the *maximum* speed, assume no air resistance, then $mg\Delta h = \frac{1}{2}mv^2$.

So $\frac{1}{2}mv^2 = 0.473...$

Rearrange to find $v = \sqrt{\frac{2 \times 0.473...}{0.72}} =$ **1.1 ms⁻¹ (to 2 s.f.)**

OR

Cancel the *m*s and rearrange to give:

$v^2 = 2g\Delta h$

$= 2 \times 9.81 \times (0.5 - 0.5 \cos 30°)$

$= 1.31429...$

$v =$ **1.1 ms⁻¹ (to 2 s.f.)**

You could be asked to apply this stuff to just about any situation in the exam. **Rollercoasters** are a bit of a favourite.

Practice Questions

Q1 State the principle of conservation of energy.

Q2 What are the equations for calculating kinetic energy and gravitational potential energy?

Q3 Show that, if there's no air resistance and the mass of the string is negligible, the speed of a pendulum is independent of the mass of the bob.

Q4 An 1800 watt kettle transfers 1000 J per second to the water inside it. The rest is lost to other forms of energy. Calculate the efficiency of the kettle.

Exam Questions

Q1 A skateboarder is on a half-pipe. He lets the board run down one side of the ramp and up the other. The height of the ramp is 2 m. Take g as 9.81 Nkg⁻¹.

a) Assume that there is no friction. Calculate his speed at the lowest point of the ramp. [3 marks]

b) How high will he rise up the other side? [1 mark]

c) Real ramps are not frictionless, so what must the skater do to reach the top on the other side? [1 mark]

Q2 A 20.0 g rubber ball is released from a height of 8.0 m. (Assume that the effect of air resistance is negligible.)

a) Find the kinetic energy of the ball just before it hits the ground. [2 marks]

b) The ball strikes the ground and rebounds to a height of 6.5 m. Calculate how much energy is converted to heat and sound in the impact with the ground. [2 marks]

Energy is never lost — it just sometimes prefers the scenic route...

Right, done, on to the next question... remember to check your answers — I can't count the number of times I've forgotten to square the velocities or to multiply by the ½. I reckon it's definitely worth the extra minute to check.

Properties of Materials

Hooke's law applies to all materials, but only up to a point. For some materials that point is so tiny you wouldn't notice...

Density is Mass per Unit Volume

1) Density is a measure of the 'compactness' (for want of a better word) of a substance. It relates the mass of a substance to how much space it takes up.

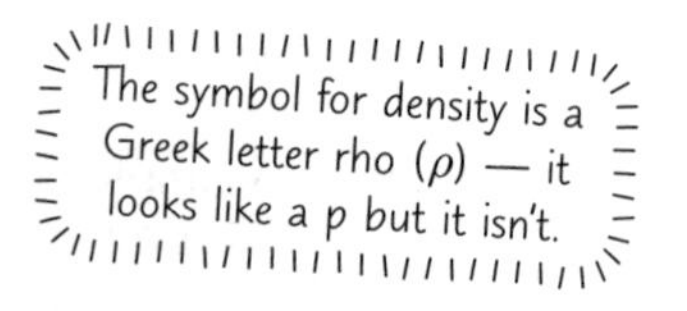

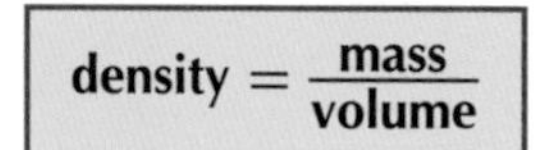

$$\rho = \frac{m}{V}$$

The **units** of **density** are **g cm^{-3}** or **kg m^{-3}**
(N.B. 1 g cm^{-3} = 1000 kg m^{-3})

2) The density of an object depends on what it's made of. Density **doesn't vary** with **size or shape**.
3) The **average density** of an object determines whether it **floats** or **sinks** — a solid object will **float** on a fluid if it has a **lower density** than the **fluid**.
4) **Water** has a density of **ρ = 1 g cm^{-3}**. So **1 cm^3** of water has a mass of **1 g**.

Hooke's Law Says that Extension is Proportional to Force

If a **metal wire** is supported at the top and then a weight is attached to the bottom, the wire **stretches**. The weight pulls down with force ***F***, producing an equal and opposite force at the support.

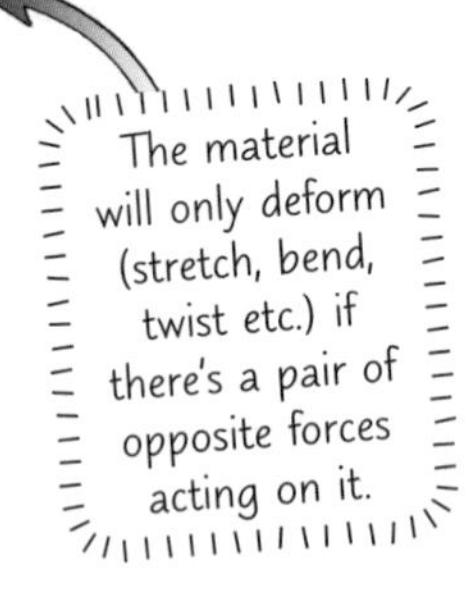

1) **Hooke's law** says that the extension of a stretched object, ***ΔL***, is proportional to the load or force, ***F***.
2) Hooke's law can be written:

$$F = k\Delta L$$

Where ***k*** is a constant (called the **stiffness constant**) that depends on the material being stretched.

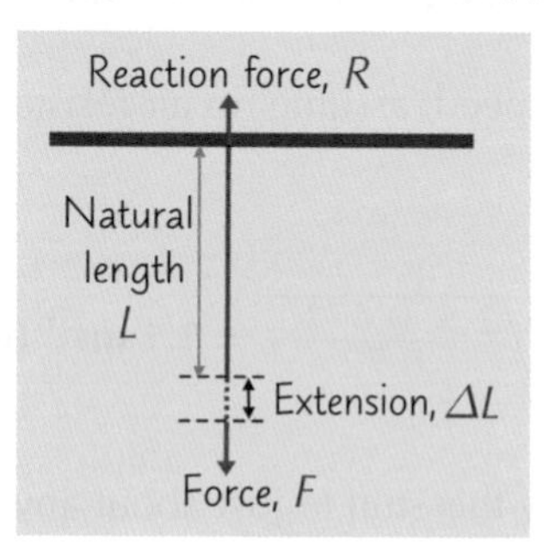

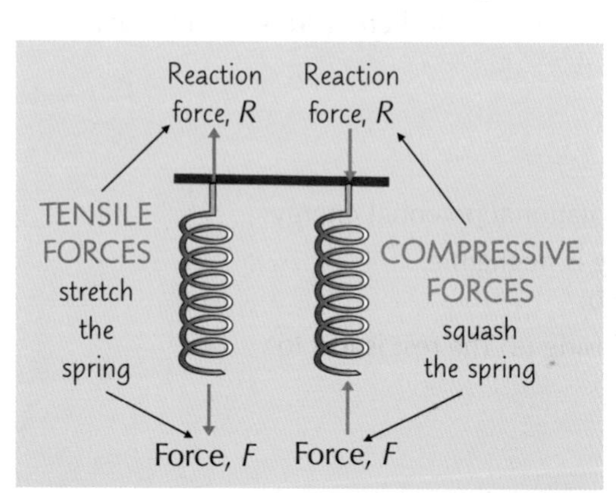

3) **Springs** obey Hooke's Law — when you apply a **pair of opposite forces,** the **extension** (or **compression**) of a spring is **proportional** to the **force** applied.
4) For springs, ***k*** is usually called the **spring constant**.
5) Hooke's law works just as well for **compressive** forces as **tensile** forces. For a spring, ***k*** has the **same value** whether the forces are tensile or compressive (that's not true for all materials).
6) **Hooke's Law** doesn't just apply to metal **springs** and **wires** — most **other materials** obey it up to a point.

Hooke's Law Stops Working when the Force is Great Enough

There's a **limit** to the force you can apply for Hooke's law to stay true.

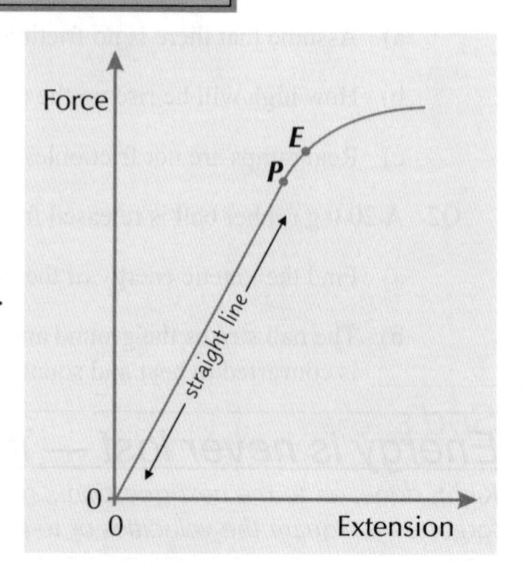

1) The graph shows force (or load) against extension for a **typical metal wire**.
2) The first part of the graph shows Hooke's law being obeyed — there's a **straight-line relationship** between **force** and **extension**.
3) When the force becomes great enough, the graph starts to **curve**. **Metals** generally obey Hooke's law up to the **limit of proportionality**, ***P***.
4) The point marked ***E*** on the graph is called the **elastic limit**. If you increase the load past the elastic limit, the material will be **permanently stretched**. When all the force is removed, the material will be **longer** than at the start.
5) Be careful — there are some materials, like **rubber**, that only obey Hooke's law for **really small** extensions.

Properties of Materials

A Stretch can be **Elastic** or **Plastic**

Elastic

If a **deformation** is **elastic**, the material returns to its **original shape** and **size** once the forces are removed.

1) When the material is put under **tension**, the **atoms** of the material are **pulled apart** from one another.
2) Atoms can **move** small distances relative to their **equilibrium positions**, without actually changing position in the material.
3) Once the **load** is **removed**, the atoms **return** to their **equilibrium** distance apart.

Elastic deformation happens as long as the **elastic limit** of the object isn't reached.

Plastic

If a deformation is **plastic**, the material is **permanently stretched**.

1) Some atoms in the material move position relative to one another.
2) When the load is removed, the **atoms don't return** to their original positions.
3) An object stretched **past its elastic limit** shows plastic deformation.

Life in plastic, it's fantastic.

Energy is Always **Conserved** When Stretching

There's more about elastic strain energy on the next two pages (plus some formulas — hurrah!).

When a material is **stretched**, **work** has to be done in stretching the material.

1) If a deformation is **elastic**, all the work done is **stored** as **elastic strain energy** in the material.
2) When the stretching force is removed, this **stored energy** is **transferred** to **other forms** — e.g. an elastic band is stretched and then fired across a room.
3) If a deformation is **plastic**, work is done to **separate atoms**, and energy is **not** stored as strain energy (it's mostly dissipated as heat).
4) This fact is used in **transport design** — **crumple zones** are designed to deform **plastically** in a **crash**. Some energy goes into **changing the shape** of the vehicle's **metal body** (and so less is transferred to the people inside).

Practice Questions

Q1 Write down the formula for calculating density. Will a material with a density of 0.8 $g\,cm^{-3}$ float on water?

Q2 State Hooke's law, and explain what is meant by the elastic limit of a material.

Q3 From studying the force-extension graph for a material as weights are suspended from it, how can you tell:
a) if Hooke's law is being obeyed, b) if the elastic limit has been reached?

Q4 What is plastic behaviour of a material under load?

Q5 Explain how crumple zones protect passengers during a car crash.

Exam Questions

Q1 A metal guitar string stretches 4.0 mm when a 10.0 N force is applied to it.

a) If the string obeys Hooke's law, calculate how far the string will stretch with a 15 N force applied to it. [1 mark]

b) Calculate the stiffness constant, k, for this string in Nm^{-1}. [2 marks]

c) The string is now stretched beyond its elastic limit. Describe what effect this will have on the string. [1 mark]

Q2 A rubber band is 6.0 cm long. When it is loaded with 2.5 N, its length becomes 10.4 cm. Further loading increases the length to 16.2 cm when the force is 5.0 N.

Does the rubber band obey Hooke's law when the force on it is 5.0 N? Justify your answer with a suitable calculation. [2 marks]

Sod's law — if you don't learn it, it'll be in the exam...

Hooke's law was discovered (unsurprisingly) by Robert Hooke 350 years ago. Three bonus facts about Mr Hooke — he was the first person to use the word 'cell' (in terms of biology, not prisons), he helped Christopher Wren with his designs for St. Paul's Cathedral, and finally no-one actually knows what he looked like. How sad. Poor old Hooke.

Stress and Strain

How much a material stretches for a particular applied force depends on its dimensions. If you want to compare it to another material, you need to use stress and strain instead. A stress-strain graph is the same for any sample of a particular material — the size of the sample doesn't matter.

A Stress Causes a Strain

A material subjected to a pair of **opposite forces** might **deform** (i.e. **change shape**).
If the forces **stretch** the material, they're **tensile**. If the forces **squash** the material, they're **compressive**.

1) **Tensile stress** is defined as the **force applied**, F, divided by the **cross-sectional area**, A:

$$\text{stress} = \frac{F}{A}$$

The **units** of stress are $\mathbf{Nm^{-2}}$ or pascals, **Pa**.

2) **Tensile strain** is defined as the **change in length** (i.e. the **extension**), divided by the **original length** of the material:

$$\text{strain} = \frac{\Delta L}{L}$$

Strain has **no units**, it's just a **ratio** and is usually written as a **number**. It can also be written as a **percentage**, e.g. extending a 0.5 m wire by 0.02 m would produce a strain of (0.02 ÷ 0.5) × 100 = 4 %.

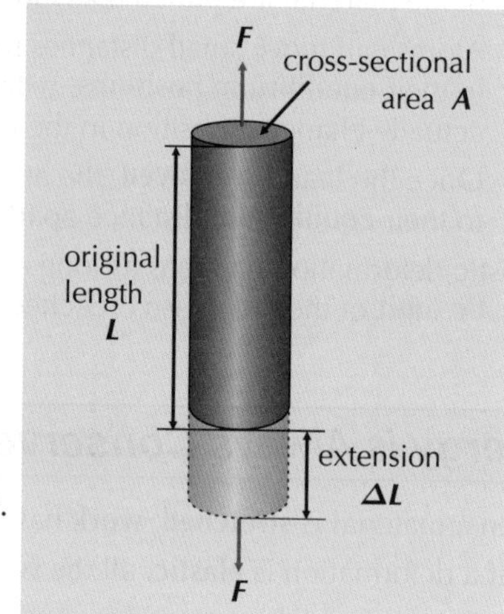

3) It doesn't matter whether the forces producing the **stress** and **strain** are **tensile** or **compressive** — the **same equations** apply. The only difference is that you tend to think of **tensile** forces as **positive**, and **compressive** forces as **negative**.

A Stress Big Enough to Break the Material is Called the Breaking Stress

As a greater and greater tensile **force** is applied to a material, the **stress** on it **increases**.

1) The effect of the **stress** is to start to **pull** the **atoms apart** from one another.
2) Eventually the stress becomes **so great** that atoms **separate completely**, and the **material breaks**. This is shown by point **B** on the graph. The stress at which this occurs is called the **breaking stress**.
3) The point marked **UTS** on the graph is called the **ultimate tensile stress**. This is the **maximum stress** that the material can withstand.
4) Both **UTS** and **B** depend on conditions e.g. **temperature**.
5) **Engineers** have to consider the **UTS** and **breaking stress** of materials when designing a **structure** — e.g. they need to make sure the stress on a material won't reach the **UTS** when the **conditions change**.

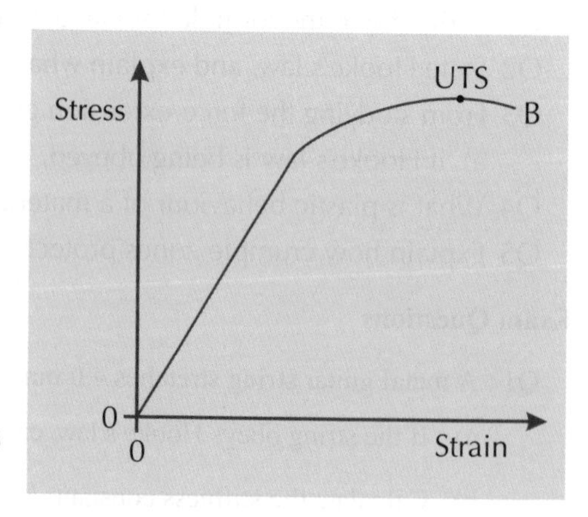

Elastic Strain Energy is the Area under a Force-Extension Graph

1) **Work** has to be done to **stretch** a material.
2) **Before** the **elastic limit** is reached, **all** this **work done** in stretching is **stored** as **elastic strain energy** in the material.
3) On a **graph** of **force against extension**, the elastic strain energy is given by the **area under the graph**.

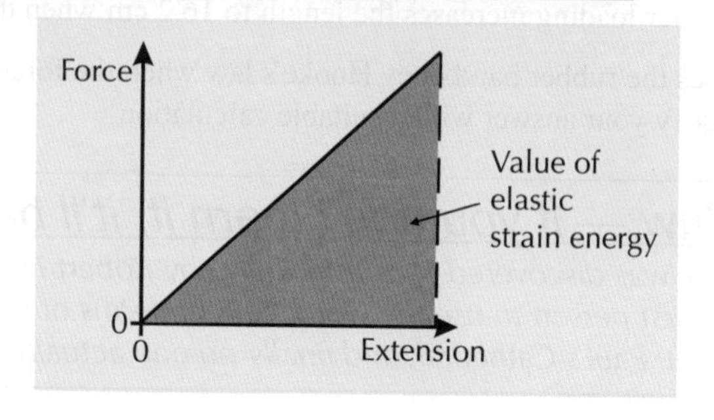

Stress and Strain

You can Calculate the **Energy Stored** in a **Stretched Wire**

Provided a material obeys Hooke's law, the **potential energy** stored inside it can be **calculated** quite easily.

1) The work done on the wire in stretching it is equal to the energy stored.
2) **Work done** equals **force × displacement**.
3) However, the **force** on the material **isn't constant**. It rises from zero up to force F. To calculate the **work done**, use the **average force** between zero and F, i.e. $\frac{1}{2}F$.

$$\text{work done} = \tfrac{1}{2}F \times \Delta L$$

This is the triangular area under the force-extension graph — see previous page.

4) Then the **elastic strain energy**, E, is: $E = \frac{1}{2}F\Delta L$
5) Because Hooke's law is being obeyed, $F = k\Delta L$, which means F can be replaced in the equation to give: $E = \frac{1}{2}k\Delta L^2$

Example: A metal wire is 55.0 cm long. A force of 550 N is applied to the wire, and the wire stretches. The length of the stretched wire is 56.5 cm. Calculate the elastic strain energy stored in the wire.

The extension of the wire is ΔL = 56.5 cm – 55.0 cm = 1.5 cm = 0.015 m

So the elastic strain energy $E = \frac{1}{2} \times F \times \Delta L$

= 1/2 × 550 × 0.015 = 4.125 J = **4.1 J (to 2 s.f.)**

Practice Questions

Q1 Write a definition for tensile stress.

Q2 Explain what is meant by the tensile strain on a material.

Q3 What is meant by the breaking stress of a material?

Q4 How can the elastic strain energy in a material under load be found from its force-extension graph?

Q5 The work done is usually calculated as force multiplied by displacement. Explain why the work done in stretching a wire is $\frac{1}{2}F\Delta L$.

Exam Questions

Q1 A steel wire is 2.00 m long. When a 300 N (to 3 s.f.) force is applied to the wire, it stretches 4.0 mm. The wire has a circular cross-section with a diameter of 1.0 mm.

a) Calculate the tensile stress in the wire. [2 marks]

b) Calculate the tensile strain of the wire. [1 mark]

Q2 A copper wire (which obeys Hooke's law) is stretched by 3.0 mm when a force of 50.0 N is applied.

a) Calculate the stiffness constant for this wire in Nm^{-1}. [2 marks]

b) Calculate the value of the elastic strain energy in the stretched wire. [1 mark]

Q3 A pinball machine contains a spring which is used to fire a small, 12.0 g metal ball to start the game. The spring has a stiffness constant of 40.8 Nm^{-1}. It is compressed by 5.00 cm and then released to fire the ball.

Calculate the maximum possible kinetic energy of the ball. [3 marks]

UTS a laugh a minute, this stuff...

Here endeth the proper physics for this section — the rest of it's materials science (and I don't care what your exam boards say). It's all a bit "useful" for my liking. Calls itself a physics course... grumble... grumble... wasn't like this in my day... But to be fair — some of it's quite interesting, and there are some lovely graphs coming up on pages 71-73.

The Young Modulus

Busy chap, Thomas Young. He did this work on tensile stress as something of a sideline. Light was his main thing. He proved that light behaved like a wave, explained how we see in colour and worked out what causes astigmatism.

The **Young Modulus** is Stress ÷ Strain

When you apply a **load** to stretch a material, it experiences a **tensile stress** and a **tensile strain**.

1) Up to the **limit of proportionality** (see p.66), the stress and strain of a material are proportional to each other.
2) So below this limit, for a particular material, stress divided by strain is a **constant**. This constant is called the **Young modulus, *E***.

$$\textbf{Young modulus} = E = \frac{\textbf{tensile stress}}{\textbf{tensile strain}} = \frac{F \div A}{\Delta L \div L} = \frac{FL}{\Delta L A}$$

Where F = force in N, A = cross-sectional area in m^2, L = initial length in m and ΔL = extension in m.

3) The **units** of the Young modulus are the same as stress (**Nm^{-2}** or pascals), since strain has no units.
4) The Young modulus is a measure of the **stiffness** of a material. It is used by **engineers** to make sure the materials they are using can withstand sufficient forces.

To **Find** the Young Modulus, You need a **Very Long Wire**

This is the experiment you're most likely to do in class:

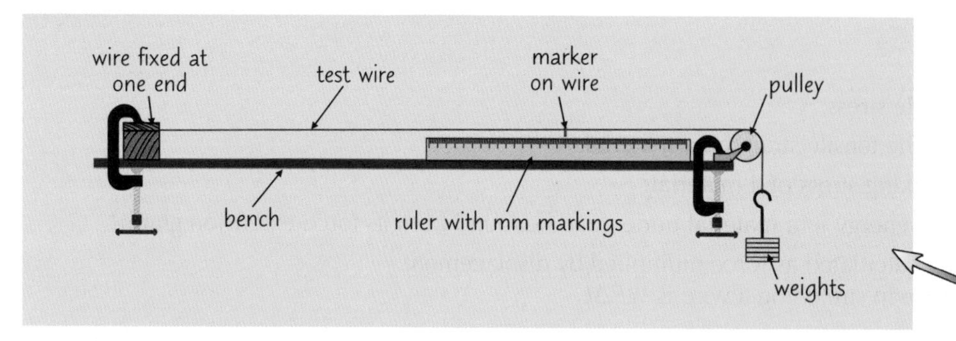

"Okay, found one. Now what?"

1) The test wire should be thin, and as long as possible. The **longer and thinner** the wire, the more it **extends** for the same force — this reduces the uncertainty in your measurements.
2) First you need to find the **cross-sectional area** of the wire. Use a **micrometer** to measure the **diameter** of the wire in several places and take an **average** of your measurements. By assuming that the cross-section is **circular**, you can use the formula for the area of a circle:

$$\textbf{area of a circle} = \pi r^2$$

Mum moment: if you're doing this experiment, wear safety goggles — if the wire snaps, it could get very messy...

3) **Clamp** the wire to the bench (as shown in the diagram above) so you can hang **weights** off one end of it. Start with the **smallest weight** necessary to **straighten** the wire. (**Don't** include this weight in your final calculations.)
4) Measure the **distance** between the **fixed end of the wire** and the **marker** — this is your unstretched length.
5) Then if you increase the weight, the **wire stretches** and the **marker moves**.
6) **Increase** the **weight** in steps (e.g. 100 g intervals), recording the marker reading each time — the **extension** is the **difference** between this reading and the **unstretched length**.
7) You can use your results from this experiment to calculate the **stress** and **strain** of the wire and plot a stress-strain curve (see next page).

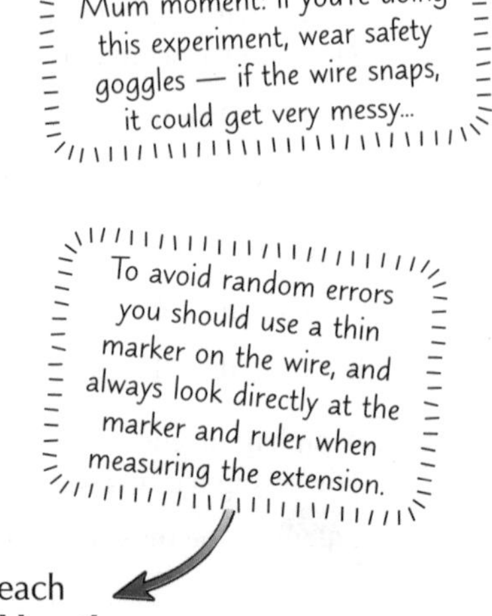

(The other standard way of measuring the Young modulus in the lab is using **Searle's apparatus**. This is a bit more accurate, but it's harder to do and the equipment's more complicated.)

The Young Modulus

Use a Stress-Strain Graph to Find E

You can plot a **graph** of **stress against strain** from your results.

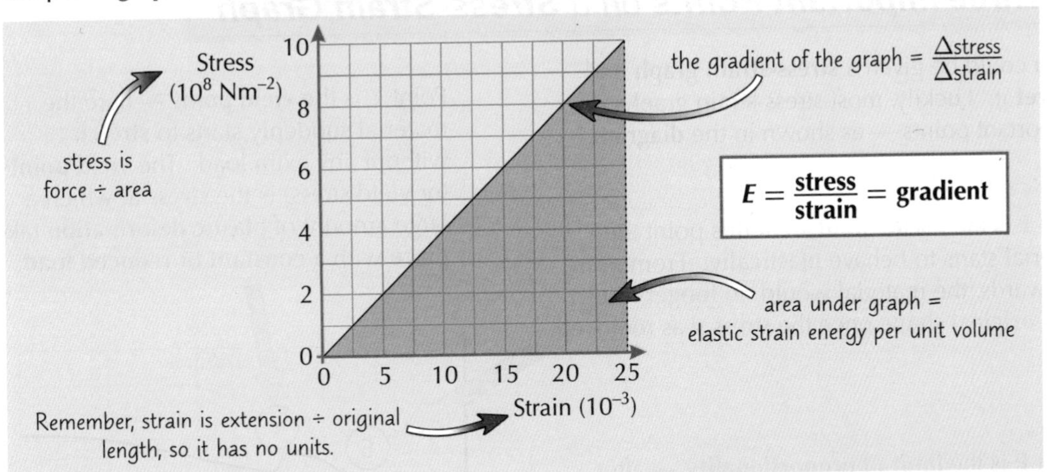

1) The **gradient** of the graph gives the Young modulus, ***E***.
2) The **area under the graph** gives the **strain energy** (or energy stored) per unit volume, i.e. the energy stored per 1 m³ of wire.
3) The stress-strain graph is a **straight line** provided that Hooke's law is obeyed, so you can also calculate the energy per unit volume as:

energy per unit vol = ½ × stress × strain

Example: The stress-strain graph above is for a thin metal wire. Find the Young modulus of the wire from the graph.

E **= change in stress ÷ change in strain = gradient**

The gradient of the graph = $\frac{\Delta stress}{\Delta strain} = \frac{10 \times 10^8}{25 \times 10^{-3}}$

$= \mathbf{4 \times 10^{10}\ Nm^{-2}}$

Practice Questions

Q1 Define the Young modulus for a material. What are the units for the Young modulus?

Q2 Describe an experiment to find the Young modulus of a test wire. Explain why a thin test wire should be used.

Q3 What is given by the area contained under a stress-strain graph?

Exam Questions

Q1 A steel wire is stretched elastically. For a load of 80.0 N, the wire extends by 3.6 mm. The original length of the wire was 2.50 m and its average diameter is 0.60 mm. Calculate the value of the Young modulus for steel. [4 marks]

Q2 Two wires, A and B, are stretched elastically under a load of 50.0 N. The original length and the extension of both wires under this load are the same. The Young modulus of wire A is found to be 7.0×10^{10} Nm^{-2}. The cross-sectional area of wire B is half that of wire A. Calculate the Young modulus of wire B. [2 marks]

Q3 The Young modulus for copper is 1.3×10^{11} Nm^{-2}.

a) The stress on a copper wire is 2.6×10^8 Nm^{-2}. Calculate the strain on the wire. [2 marks]

b) The load applied to the copper wire is 100 N (to 3 s.f.). Calculate the cross-sectional area of the wire. [2 marks]

c) Calculate the strain energy per unit volume for this loaded wire. [2 marks]

Learn that experiment — it's important...

Getting back to the good Dr Young... As if ground-breaking work in light, the physics of vision and materials science wasn't enough, he was also a well-respected physician, a linguist and an Egyptologist. He was one of the first to try to decipher the Rosetta stone (he didn't get it right, but nobody's perfect). Makes you feel kind of inferior, doesn't it...

Stress-Strain and Force-Extension Graphs

I hope the stresses and strains of this section aren't getting to you too much.
Don't worry, though — there's just these two pages to go before you're on to the electrifying world of electricity.

There are **Three Important Points** on a **Stress-Strain Graph**

In the exam you could be given a **stress-strain graph** and asked to **interpret** it. Luckily, most stress-strain graphs share **three** important points — as shown in the **diagram**.

Point **Y** is the **yield point** — here the material suddenly starts to **stretch** without any extra load. The **yield point** (or yield stress) is the **stress** at which a large amount of **plastic deformation** takes place with a **constant** or **reduced load**.

Point **E** is the **elastic limit** — at this point the material starts to behave **plastically**. From point E onwards, the material would **no longer** return to its **original shape** once the stress was removed.

Point **P** is the **limit of proportionality** — after this, the graph is no longer a straight line but starts to **bend**. At this point, the material **stops** obeying **Hooke's law**, but would still **return** to its **original shape** if the stress was removed.

Before point **P**, the graph is a **straight line** through the **origin**. This shows that the material is obeying **Hooke's law** (page 66). The **gradient** of the line is constant — it's the **Young modulus** (see pages 70-71).

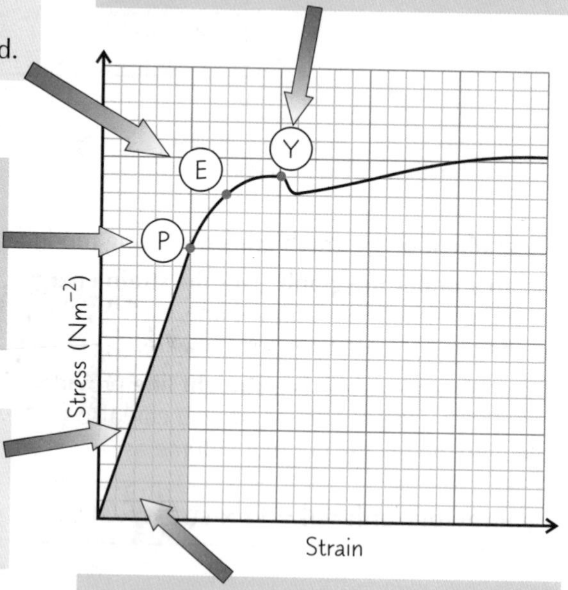

The **area** under the first part of the graph gives the **energy stored** in the **material per unit volume** (see page 71).

Stress-Strain Graphs for **Brittle** Materials **Don't Curve**

The graph shown below is typical of a **brittle** material.

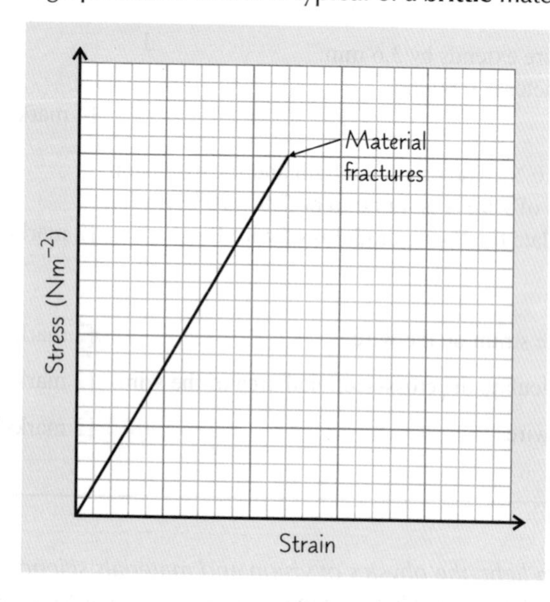

1) The graph starts the same as the one above — with a **straight line through the origin**. So brittle materials also obey **Hooke's law**.
2) However, when the **stress** reaches a certain point, the material suddenly **fractures** (breaks) — it **doesn't deform plastically**.
3) A **chocolate bar** is an example of a brittle material — you can break chunks of chocolate off the bar without the whole thing changing shape.
4) **Ceramics** (e.g. **glass** and **pottery**) are brittle too — they tend to shatter.

Stress-Strain and Force-Extension Graphs

Force-Extension Graphs Are Similar to *Stress-Strain* Graphs

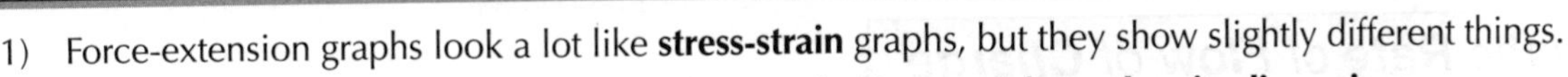

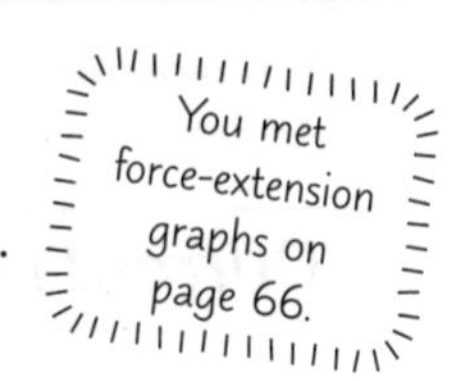

1) Force-extension graphs look a lot like **stress-strain** graphs, but they show slightly different things.
2) Force-extension graphs are **specific** for the tested **object** and **depend on its dimensions**. Stress-strain graphs describe the **general behaviour** of a **material**, because stress and strain are **independent of the dimensions**.
3) You can plot a force-extension graph of what happens when you gradually **remove** a **force** from an object. The **unloading** line doesn't always match up with the **loading** line though.

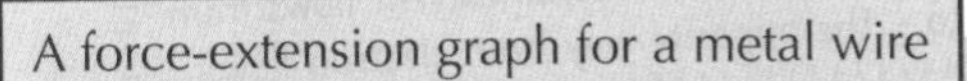
A force-extension graph for a metal wire

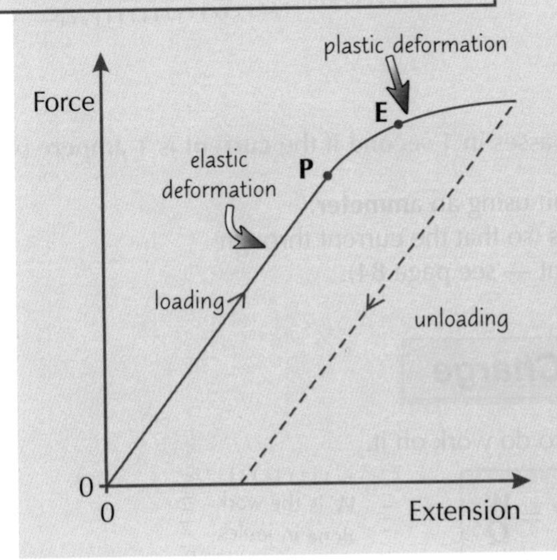

1) This graph is for a metal wire that has been stretched beyond its **limit of proportionality (P)** so the graph starts to **curve**.
2) When the load is **removed**, the **extension decreases**.
3) The unloading line is **parallel** to the loading line because the stiffness constant ***k*** is still the same (since the forces between the atoms are the same as they were during the loading).
4) But because the wire was stretched beyond its **elastic limit (E)** and deformed **plastically**, it has been **permanently stretched**. This means the unloading line doesn't go through the origin.
5) The **area** between the two lines is the **work done** to permanently deform the wire.

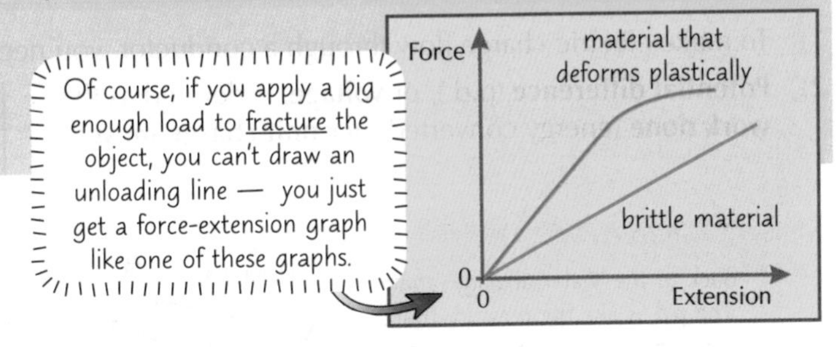

Practice Questions

Q1 What is the difference between the limit of proportionality and the elastic limit?

Q2 Describe what happens at the yield point.

Q3 A metal wire is stretched beyond its elastic limit. Why does the unloading line on the force-extension graph for the wire not go through the origin?

Exam Questions

Q1 Hardened steel is a hard, brittle form of steel made by heating it up slowly and then quenching it in cold water.

a) State what is meant by the term brittle. [1 mark]

b) Sketch a stress-strain graph for hardened steel. [2 marks]

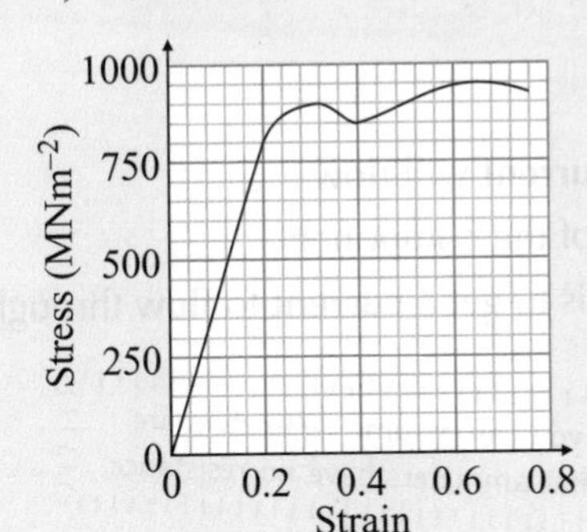

Q2 The diagram shows a stress-strain graph for a nylon thread.

a) State the yield stress for nylon. [1 mark]

b) Calculate how much energy per unit volume is stored in the thread when the limit of proportionality is reached. [2 marks]

c) The unloading curve is added to the stress-strain graph. Describe how the work done per unit volume to permanently deform the thread can be calculated. [2 marks]

My sister must be brittle — she's always snapping...

It's the end of the section on materials, but don't shed a tear — there's some good stuff coming up in the next one. You can always come back and re-read these pages too — it's a good idea, even if you could do all of the practice questions.

Current, Potential Difference and Resistance

You wouldn't reckon there was that much to know about electricity... just plug something in, and bosh — electricity. Ah well, never mind the age of innocence — here are all the gory details...

Current is the Rate of Flow of Charge

1) The **current** in a **wire** is like **water** flowing in a **pipe**. The **amount** of water that flows depends on the **flow rate** and the **time**. It's the same with electricity — **current is the rate of flow of charge**.

$$\Delta Q = I\Delta t \text{ or } I = \frac{\Delta Q}{\Delta t}$$

Where ΔQ is the charge in coulombs, I is the current in amperes and Δt is the time taken in seconds.

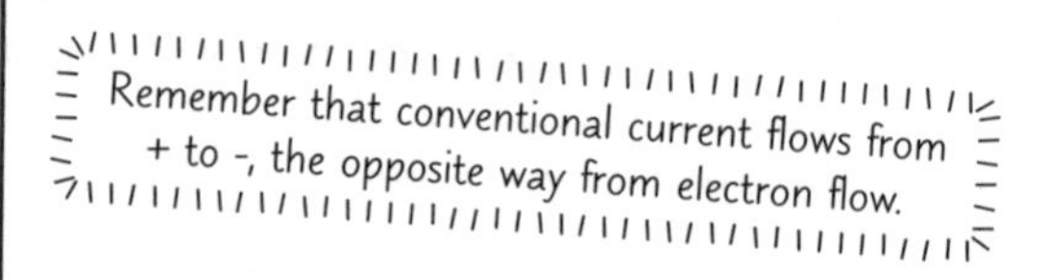

2) The **coulomb** is the unit of **charge**.

One **coulomb** (**C**) is defined as the **amount of charge** that passes in **1 second** if the **current** is **1 ampere** (**A**).

3) You can measure the current flowing through a part of a circuit using an **ammeter**. Remember — you always need to attach an ammeter in **series** (so that the current through the ammeter is the same as the current through the component — see page 84).

Potential Difference is the Energy per Unit Charge

1) To make electric charge flow through a conductor, you need to do work on it.
2) **Potential difference** (p.d.), or **voltage**, is defined as the **work done** (energy converted) **per unit charge moved**.

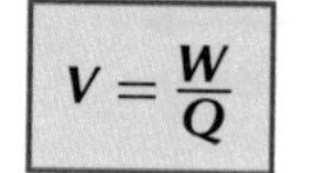

W is the work done in joules.

Back to the 'water analogy' again. The p.d. is like the pressure that's forcing water along the pipe.

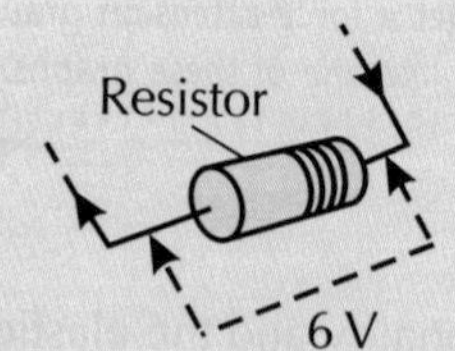

Here you do 6 J of work moving each coulomb of charge through the resistor, so the p.d. across it is 6 V. The energy gets converted to heat.

3) You can measure the potential difference across a component using a **voltmeter**.
4) Remember, the potential difference across components in parallel is **the same**, so the **voltmeter** should be connected in **parallel** with the component.

Definition of the Volt

The **potential difference** across a component is **1 volt** when you convert **1 joule** of energy moving **1 coulomb** of charge through the component.

$$1\text{ V} = 1\text{ JC}^{-1}$$

Everything has Resistance

1) If you put a **potential difference** (p.d.) across an **electrical component**, a **current** will flow.
2) **How much** current you get for a particular **p.d.** depends on the **resistance** of the component.
3) You can think of a component's **resistance** as a **measure** of how **difficult** it is to get a **current** to **flow** through it.

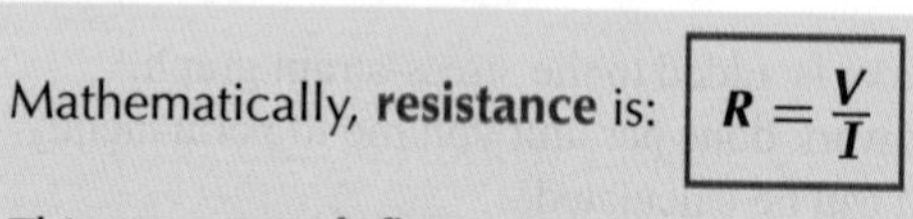

This equation **defines** resistance.

Unless told otherwise, you can assume voltmeters are infinitely resistant, and that ammeters have no resistance.

4) **Resistance** is measured in **ohms** (Ω).

A component has a resistance of 1 Ω if a **potential difference** of **1 V** makes a **current** of **1 A** flow through it.

Current, Potential Difference and Resistance

For an **Ohmic Conductor**, **R** is a **Constant**

1) A chap called **Ohm** did most of the early work on resistance. He developed a rule to **predict** how the **current** would **change** as the applied **potential difference increased**, for **certain types** of conductor.
2) The rule is now called **Ohm's law** and the conductors that **obey** it (mostly metals) are called **ohmic conductors**.

Provided the **physical conditions**, such as **temperature**, remain **constant**, the **current** through an ohmic conductor is **directly proportional** to the **potential difference** across it.

$I \propto V$

1) This graph shows what happens if you plot current against voltage for an ohmic conductor.
2) As you can see it's a **straight-line** graph — **doubling** the **p.d. doubles** the **current**.
3) What this means is that the **resistance** is **constant** — $V \div I$ is always a fixed value.
4) Often **factors** such as **light level** or **temperature** will have a **significant effect** on resistance (the resistivity changes, see page 78), so you need to remember that Ohm's law is **only** true for **ohmic conductors** under **constant physical conditions**, e.g. temperature.
5) Ohm's law is a **special case** — lots of components aren't ohmic conductors and have characteristic current-voltage (I–V) graphs of their very own (see pages 76-77).

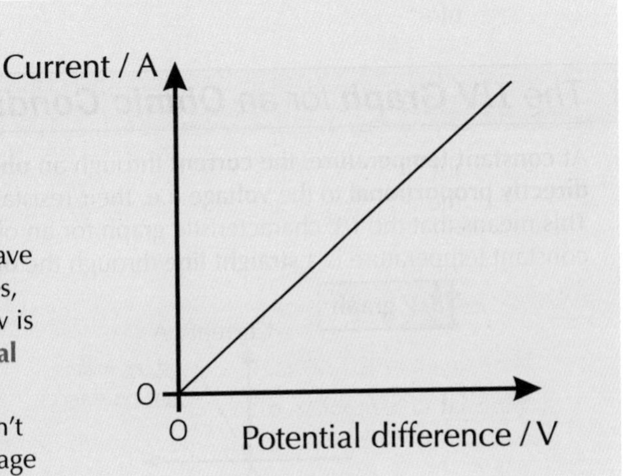

Practice Questions

Q1 Describe in words how current and charge are related.

Q2 Define the coulomb.

Q3 Define potential difference.

Q4 Give the values of resistance that we assume voltmeters and ammeters to have.

Q5 Name one environmental factor likely to alter the resistance of a component.

Q6 What is special about an ohmic conductor?

Exam Questions

Q1 A battery delivers 4500 C of electric charge to a circuit in 10.0 minutes. Calculate the average current. [1 mark]

Q2 An electric motor runs off a 12 V d.c. supply.
Calculate how much electric charge will pass through the motor when it transfers 120 J of energy. [2 marks]

Q3 A current of 12 amps flows through an ohmic conductor when a potential difference of 2.0 V is applied across it. Assume the temperature of the conductor remains constant.

a) Calculate the resistance of the conductor. [1 mark]

b) Calculate the current through the conductor when the potential difference across it is 35 V. [1 mark]

c) Sketch the I-V graph for the conductor when potential differences of up to 35 V are applied to it. [1 mark]

I can't even be bothered to make the current joke...

Talking of currant jokes, I saw this bottle of wine the other day called 'raisin d'être' — 'raison d'être' of course meaning 'reason for living', but spelled slightly different to make 'raisin', meaning 'grape'. Ho ho. Chuckled all the way home.

I/V Characteristics

Woohoo — real physics. This stuff's actually kind of interesting.

I/V Graphs Show how Resistance Varies

1) The term '***I/V* characteristic**' refers to a **graph** of I **against** V which shows how the **current** (I) flowing through a **component changes** as the **potential difference** (V) across it is increased.
2) This **diagram** shows the type of **circuit** used to obtain a characteristic I/V graph for a component.

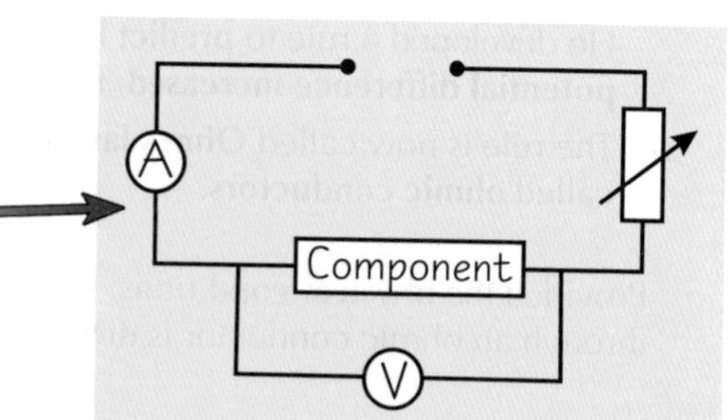

You could also be asked about a ***V/I* graphs**. They're pretty similar, but with V plotted against I. The **resistance** at a point on the graph is simply V/I at that point.

The I/V Graph for an Ohmic Conductor is a Straight Line through the Origin

At **constant temperature**, the **current** through an **ohmic conductor** (e.g. metals) is **directly proportional** to the **voltage** (i.e. their resistance is constant, see page 75). This means that the I/V characteristic graph for an ohmic conductor at a constant temperature is a **straight line** through the **origin**.

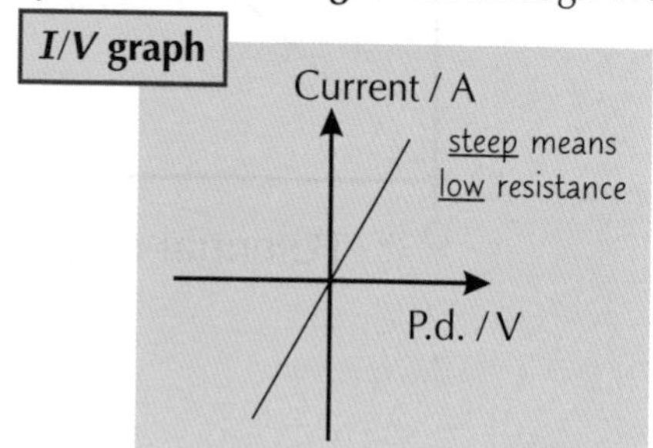

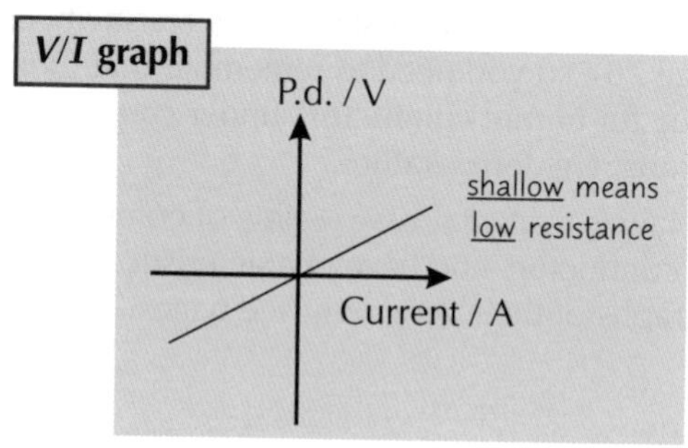

The I/V Characteristic for a Filament Lamp is a Curve

The I/V characteristic for a **filament lamp** is a **curve** that starts **steep** but gets **shallower** as the **voltage rises**.

The **filament** in a lamp is just a **coiled-up** length of **metal wire**, so you might think it should have the **same characteristic graph** as a **metallic conductor**. It doesn't because it **gets hot**. **Current** flowing through the lamp **increases** its **temperature**.

The **resistance** of a **metal increases** as the **temperature increases**.

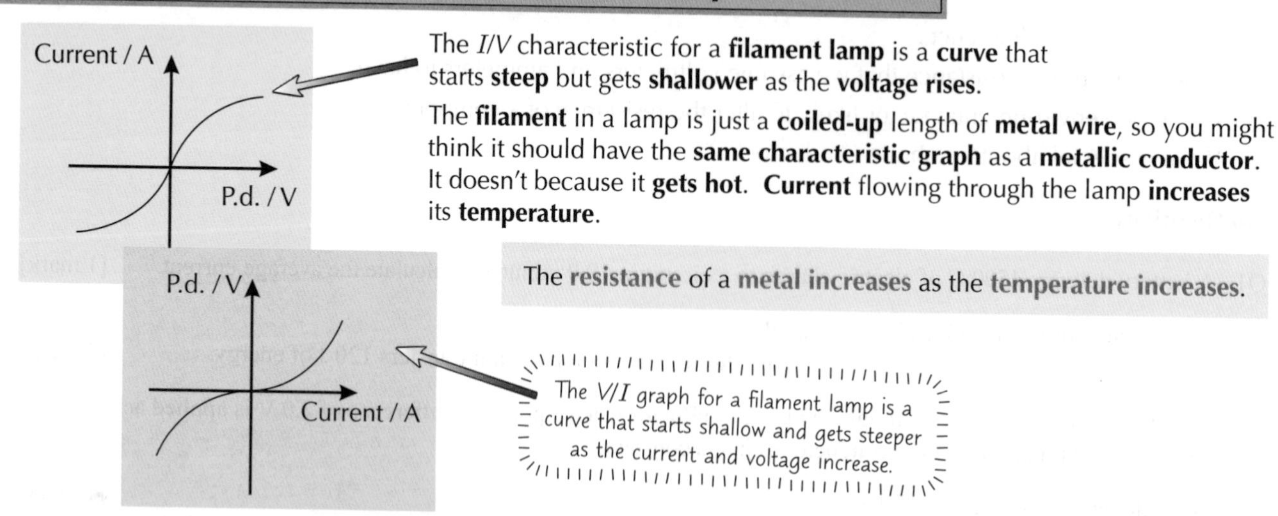

Semiconductors are Used in Sensors

Semiconductors are **nowhere near** as good at **conducting** electricity as **metals**. This is because there are far, far **fewer charge carriers** available. However, if **energy** is supplied to the semiconductor, **more charge carriers** can be **released**. This means that they make **excellent sensors** for detecting **changes** in their **environment**.

You need to know about the **two** semiconductor components on the next page — **thermistors** and **diodes**.

I/V Characteristics

The Resistance of a Thermistor Depends on Temperature

1) A **thermistor** is a **resistor** with a **resistance** that depends on its **temperature**.
2) You only need to know about **NTC** thermistors — NTC stands for 'Negative Temperature Coefficient'.
3) This means that the **resistance decreases** as the **temperature goes up**.

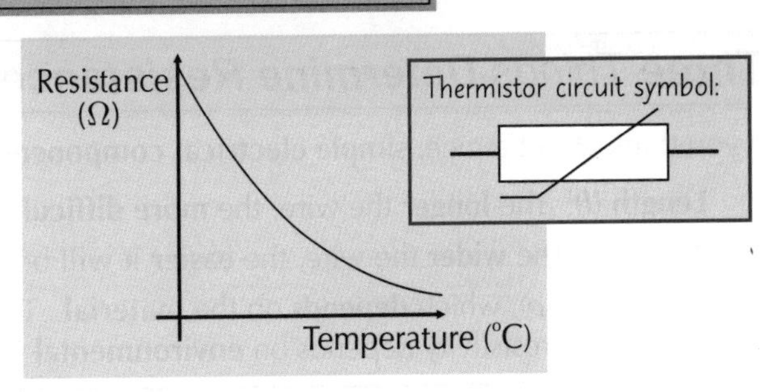

The *I/V* characteristic graph for an NTC thermistor curves upwards.

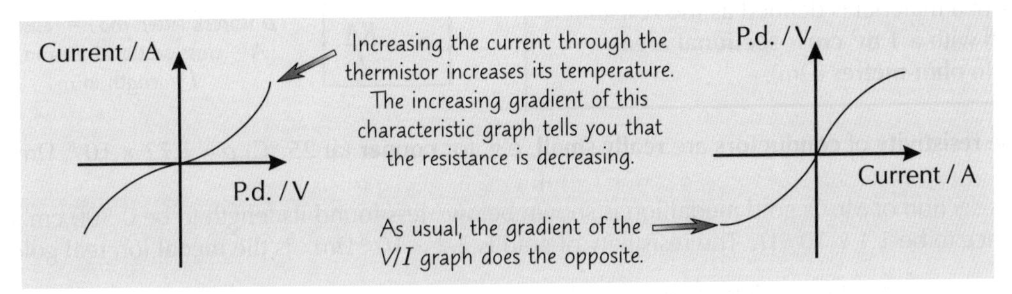

Warming the thermistor gives more **electrons** enough **energy** to **escape** from their atoms. This means that there are **more charge carriers** available, so the resistance is lower.

Diodes Only Let Current Flow in One Direction

Diodes (including light emitting diodes (LEDs)) are designed to let **current flow** in **one direction** only. You don't need to be able to explain how they work, just what they do.

Diode and LED circuit symbols:

diode LED

1) **Forward bias** is the **direction** in which the **current** is **allowed to flow**.
2) **Most** diodes require a **threshold voltage** of about **0.6 V** in the **forward direction** before they will conduct.
3) In **reverse bias**, the **resistance** of the diode is **very high** and the current that flows is **very tiny**.

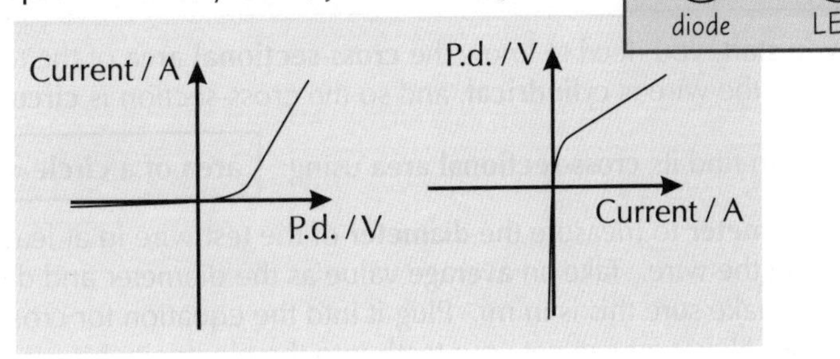

Practice Questions

Q1 Sketch the circuit used to determine the *I/V* characteristics of a component.

Q2 Why does an *I/V* graph curve for a filament lamp?

Q3 Draw an *I/V* characteristic graph for a diode.

Exam Questions

Q1 a) Describe the shape of a current-voltage characteristic graph for a filament lamp. [1 mark]

b) Explain the changes in the gradient of the graph as the current increases. [2 marks]

Q2 Explain, with reference to charge carriers, how an NTC thermistor connected in a circuit can be used as a temperature sensor. [3 marks]

You light up my world like an LED — with One Directional current...

Learn the graphs on these two pages, and check that you can explain them. Remember that a temperature increase causes an increase in resistance in a filament lamp, but a decrease in resistance in a thermistor.

Resistivity and Superconductivity

Resistance is all well and good, but it depends on the size of the thing doing the resisting. If you want to compare two materials, you need a quantity that doesn't depend on size. Enter 'resistivity'...

Three Things Determine Resistance

If you think about a nice, **simple electrical component**, like a **length of wire**, its **resistance** depends on:

1) **Length** (l). The **longer** the wire, the **more difficult** it is to make a **current flow**.
2) **Area** (A). The **wider** the wire, the **easier** it will be for the electrons to pass along it.
3) **Resistivity** (ρ), which **depends** on the **material**. The **structure** may make it easy or difficult for charge to flow. In general, resistivity depends on **environmental factors** as well, like **temperature** and **light intensity**.

The **resistivity** of a material is defined as the **resistance** of a **1 m length** with a **1 m² cross-sectional area**. It is measured in **ohm-metres** (Ωm).

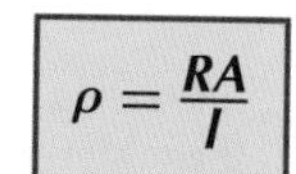

$$\rho = \frac{RA}{l}$$

ρ (Greek letter 'rho') = resistivity, A = cross-sectional area in m², l = length in m

Typical values for the **resistivity** of **conductors** are **really small**, e.g. for **copper** (at 25 °C) $\rho = 1.72 \times 10^{-8}$ Ωm.

Example: A cross-section of Mr T's gold medallion is shown below. Jess found its length to be 0.500 cm and its resistance to be 1.1×10^{-8} Ω. The resistivity of gold is 2.2×10^{-8} Ωm. Is the medallion real gold?

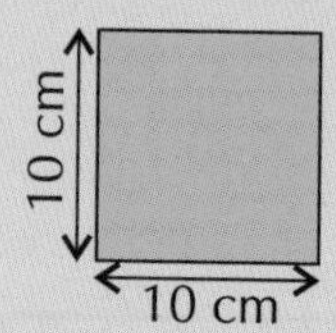

Convert all the lengths into metres.

The cross-sectional area of the medallion = 0.1 m × 0.1 m = 0.01 m². Length = l = 0.005 m

So the resistivity = $\rho = \frac{RA}{l} = \frac{(1.1 \times 10^{-8} \times 0.01)}{0.005} = \mathbf{2.2 \times 10^{-8}\,\Omega m}$

So Mr T's gold is the real deal.

To Find the Resistivity of a Wire you Need to Find its Resistance

Before you start, you need to know the **cross-sectional area** of the test wire. Assume that the wire is **cylindrical**, and so the cross-section is **circular**.

Then you can find its **cross-sectional area** using: **area of a circle = πr^2**

Use a **micrometer** to measure the **diameter** of the test wire in at least **three** different points along the wire. Take an **average** value as the diameter and divide by **2** to get the **radius** (make sure this is in m). Plug it into the equation for cross-sectional area and... **ta da**. Now you can get your teeth into the electricity bit...

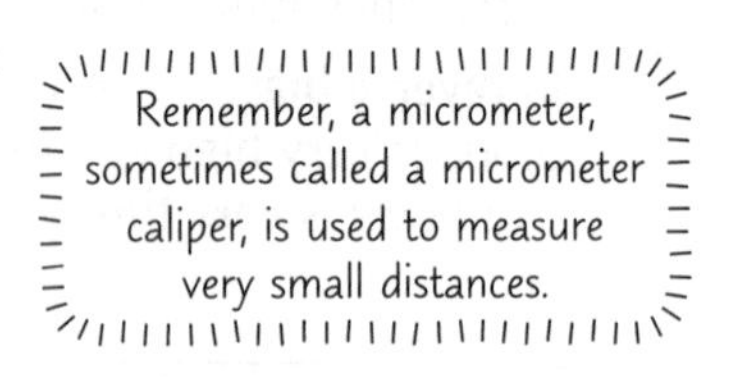

1) The **test wire** should be **clamped** to a ruler with the circuit attached to the wire where the ruler reads zero.
2) Attach the **flying lead** to the test wire — the lead is just a wire with a crocodile clip at the end to allow connection to any point along the test wire.
3) Record the **length** of the test wire **connected** in the circuit, the **voltmeter reading** and the **ammeter reading**.
4) Use your readings to calculate the **resistance** of the length of wire, using: $R = \frac{V}{I}$
5) Repeat this measurement and calculate an average resistance for the length.
6) Repeat for several **different** lengths, for example between 0.10 and 1.00 m.
7) Plot your results on a graph of **resistance** against **length**, and draw a **line of best fit** (see page 91).

The **gradient** of the line of best fit is equal to $\frac{R}{l} = \frac{\rho}{A}$. So **multiply** the **gradient** of the line of best fit by the **cross-sectional area** of the wire to find the resistivity of the wire material.

8) The **resistivity** of a material depends on its **temperature**, so you can only find the resistivity of a material **at a certain temperature**. Current flowing in the test wire can cause its temperature to increase, which can lead to random errors and invalid results (see p.89). Try to keep the temperature of the test wire **constant**, e.g. by only having small currents flow through the wire.

Resistivity and Superconductivity

Superconductors Have **Zero Resistivity**

I couldn't find a conductor, so you'll have to make do with this instead.

1) Normally, all materials have **some resistivity** — even really good conductors like silver and copper.
2) That resistance means that whenever electricity flows through them, they **heat up**, and some of the electrical energy is **wasted** as thermal energy (heat).
3) You can **lower** the resistivity of many materials like metals by **cooling them down**.
4) If you **cool** some materials (e.g. mercury) down to below a **'transition temperature'**, their **resistivity disappears entirely** and they become a **superconductor**.
5) Without any resistance, **none** of the electrical energy is turned into heat, so **none** of it's wasted. That means you can start a current flowing in a circuit using a magnetic field, take away the magnet and the current would carry on flowing **forever**. Neat.
6) There's a catch, though. Most 'normal' conductors, e.g. metals, have transition temperatures below **10 kelvin** (**–263 °C**). Getting things that cold is **hard**, and **really expensive**.
7) Solid-state physicists all over the world are trying to develop **room-temperature superconductors**. So far, they've managed to get some weird **metal oxide** things to superconduct at about **140 K** (**–133 °C**), which is a much easier temperature to get down to. They've still got a long way to go though.

Uses of Superconductors

Using superconducting wires you could make:

1) **Power cables** that transmit electricity without any **loss** of power.
2) Really **strong electromagnets** that **don't** need a constant power source (for use in medical applications and Maglev trains).
3) **Electronic circuits** that work really **fast**, because there's no resistance to slow them down.

Practice Questions

Q1 What three factors does the resistance of a length of wire depend on?

Q2 What are the units for resistivity?

Q3 What happens to mercury when it's cooled to its transition temperature?

Exam Questions

Q1 Aluminium has a resistivity of 2.8×10^{-8} Ωm at 20 °C and a transition temperature of 1.2 K.

a) Calculate the resistance of a pure aluminium wire of length 4.00 m and diameter 1.0 mm, at 20 °C. [3 marks]

b) The wire is cooled to a temperature of 1 K. What is its resistance now? Explain your answer. [2 marks]

Q2 A student is trying to identify a piece of unknown thin metal wire using the table of resistivities of common metals below. She measures the potential difference across and current through different length pieces of the wire and calculates the resistance of each length of wire.

Metal	Resistivity at 20°C
Aluminium	2.82×10^{-8} Ωm
Silver	1.59×10^{-8} Ωm
Tungsten	5.6×10^{-8} Ωm

a) Explain why she must keep the temperature of the wire constant at 20°C. [2 marks]

b) State one further measurement she must make and suggest an appropriate measuring instrument. [2 marks]

c) State one assumption that she must make about the wire in order to calculate the resisitivity. [1 mark]

Superconductors and Johnny Depp — just too cool to resist...

Superconducting electromagnets are used in magnetic resonance imaging (MRI) scanners in hospitals. That way, the huge magnetic fields they need can be generated without using up a load of electricity. Great stuff...

Electrical Energy and Power

Power and energy are pretty familiar concepts — and here they are again. Same principles, just different equations.

Power is the Rate of Transfer of Energy

Power (***P***) is **defined** as the **rate** of **transfer** of **energy**.
It's measured in **watts** (***W***), where **1 watt** is equivalent to **1 joule per second**.

or $P = \frac{E}{t}$

There's a really simple formula for **power** in **electrical circuits**:

$P = VI$

This makes sense, since:

1) **Potential difference** (V) is defined as the **energy transferred** per **coulomb**.
2) **Current** (I) is defined as the **number** of **coulombs** transferred per **second**.
3) So **p.d.** × **current** is **energy transferred per second**, i.e. **power**.

He didn't know when, he didn't know where... but one day this PEt would get his revenge.

You know from the definition of **resistance** that: $V = IR$

Combining the **two equations** gives you loads of **different ways** to **calculate power**.

$P = VI$ $\qquad P = \frac{V^2}{R}$ $\qquad P = I^2R$

Obviously, which equation you should use depends on what **quantities** you're given in the **question**.

Phew... that's quite a few equations to learn and love. And as if they're not exciting enough, here are some examples to get your teeth into...

Example 1: A 24 W car headlamp is connected to a 12 V car battery.
Assume the wires connecting the lamp to the battery have negligible resistance.
a) How much energy will the lamp convert into light and heat energy in 2 hours?
b) Find the total resistance of the lamp.

a) Number of seconds in 2 hours = 2 × 60 × 60 = 7200 s
$E = P \times t = 24 \times 7200 = 172\ 800$ J = **170 kJ (to 2 s.f.)**

b) Rearrange the equation $P = \frac{V^2}{R}$, $R = \frac{V^2}{P} = \frac{12^2}{24} = \frac{144}{24} = \mathbf{6\ \Omega}$

Example 2: A robotic mutant Santa from the future converts 750 J of electrical energy into heat every second.
a) What is the power of the robotic mutant Santa?
b) All of the robotic mutant Santa's components are connected in series, with a total resistance of 30 Ω. What current flows through his wire veins?

a) Power = $\frac{E}{t} = \frac{750}{1} = \mathbf{750\ W}$

b) Rearrange the equation $P = I^2R$, $I = \sqrt{\frac{P}{R}} = \sqrt{\frac{750}{30}} = \sqrt{25} = \mathbf{5\ A}$

Electrical Energy and Power

Energy is Easy to **Calculate** if you Know the **Power**

Sometimes it's the **total energy** transferred that you're interested in. In this case you simply need to **multiply** the **power** by the **time**. So:

$E = VIt$ (or $E = \frac{V^2}{R}t$, or $E = I^2Rt$)

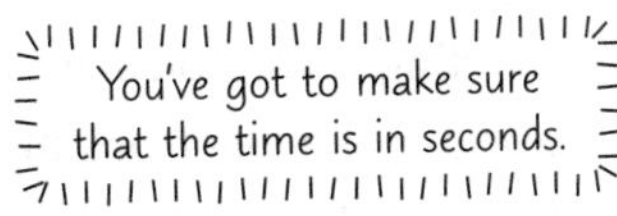

Example:

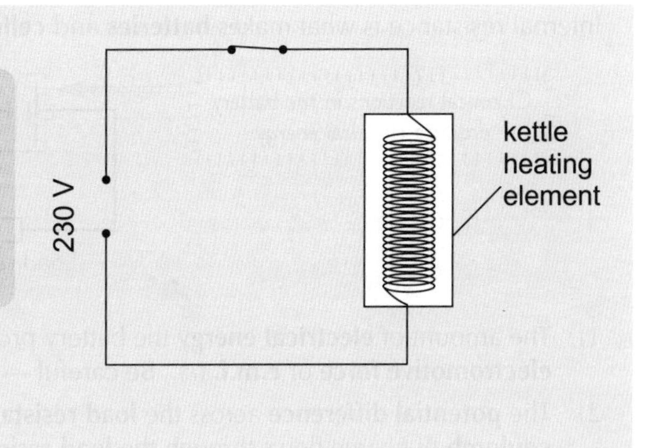

Betty pops the kettle on to make a brew.
It takes 4.5 minutes for the kettle to boil the water inside it.
A current of 4.0 A flows through the kettle's heating element once it is connected to the mains (230 V).
How much energy does the kettle's heating element transfer to the water in the time it takes to boil?

Time the kettle takes to boil in seconds = 4.5 × 60 = 270 s.
You have the current, potential difference, and time taken, so use the equation $E = VIt$:
$E = 230 \times 4.0 \times 270 = 248\,400$ J = **250 kJ (to 2.s.f.)**

Practice Questions

Q1 Power is measured in watts. What is 1 watt equivalent to?

Q2 What equation links power, voltage and resistance?

Q3 Write down the equation linking power, current and resistance.

Exam Questions

Q1 The circuit diagram for a mains-powered hair dryer is shown below.

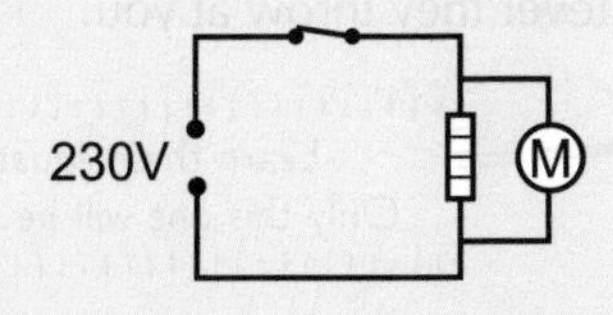

a) The heater has a power of 920 W in normal operation. Calculate the current in the heater. [1 mark]

b) The motor has a resistance of 190 Ω.
What current will flow through the motor when the hair dryer is in use? [1 mark]

c) Show that the total power of the hair dryer in normal operation is about 1.2 kW. [2 marks]

Q2 A 12 V car battery supplies a current of 48 A for 2.0 seconds to the car's starter motor.
The total resistance of the connecting wires is 0.01 Ω.

a) Calculate the energy transferred from the battery. [2 marks]

b) Calculate the energy wasted as heat in the wires. [2 marks]

Ultimate cosmic powers...

Whenever you get equations in this book, you know you're gonna have to learn them. Fact of life. I used to find it helped to stick big lists of equations all over my walls in the run-up to the exams. But as that's possibly the least cool wallpaper imaginable, I don't advise inviting your friends round till after the exams...

E.m.f. and Internal Resistance

There's resistance everywhere — inside batteries, in all the wires (although it's very small) and in the components themselves. I'm assuming the resistance of the wires is zero on the next two pages, but you can't always do this.

Batteries have ***Resistance***

Remember, I'm assuming that the resistance of the wires in the circuit is zero.

Resistance comes from **electrons colliding** with **atoms** and **losing energy** to other forms.
In a **battery**, **chemical energy** is used to make **electrons move**. As they move, they collide with atoms inside the battery — so batteries **must** have resistance. This is called **internal resistance**.
Internal resistance is what makes **batteries** and **cells warm up** when they're used.

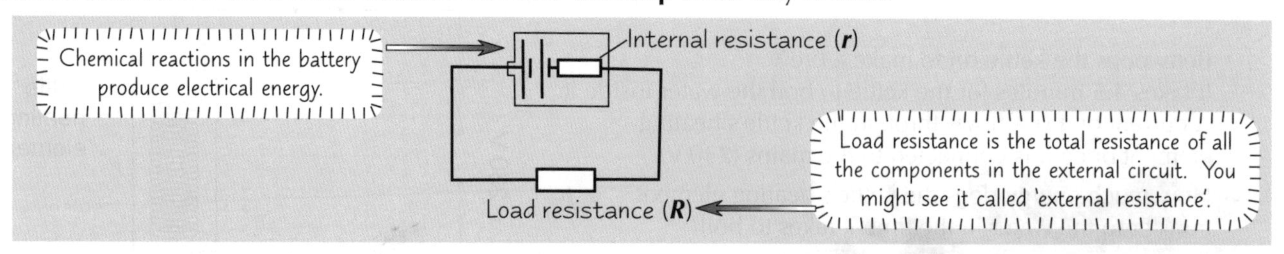

1) The amount of **electrical energy** the battery produces for each **coulomb** of charge is called its **electromotive force** or **e.m.f.** (ε). Be careful — e.m.f. **isn't** actually a force. It's measured in **volts**.

$$\varepsilon = \frac{E}{Q}$$

2) The **potential difference** across the **load resistance** (***R***) is the **energy transferred** when **one coulomb** of charge flows through the **load resistance**. This potential difference is called the **terminal p.d.** (***V***).
3) If there was **no internal resistance**, the **terminal p.d.** would be the **same** as the **e.m.f.** However, in **real** power supplies, there's **always some energy lost** overcoming the internal resistance.
4) The **energy wasted per coulomb** overcoming the internal resistance is called the **lost volts** (***v***).

Conservation of energy tells us:

energy per coulomb supplied by the source = **energy per coulomb transferred in load resistance** + **energy per coulomb wasted in internal resistance**

There are Loads of ***Calculations*** *with* ***E.m.f.*** *and* ***Internal Resistance***

Examiners can ask you to do **calculations** with **e.m.f.** and **internal resistance** in loads of **different** ways. You've got to be ready for whatever they throw at you.

$$\varepsilon = V + v \qquad \varepsilon = I(R + r)$$
$$V = \varepsilon - v \qquad V = \varepsilon - Ir$$

Learn these equations for the exam. Only this one will be on your formula sheet.

These are all basically the **same equation**, just written differently. If you're given enough information you can calculate the e.m.f. (ε), terminal p.d. (***V***), lost volts (***v***), current (***I***), load resistance (***R***) or internal resistance (***r***). Which equation you should use depends on what information you've got, and what you need to calculate.

You Can Work Out the ***E.m.f.*** *of* ***Multiple*** *Cells in* ***Series*** *or* ***Parallel***

For cells **in series** in a circuit, you can calculate the **total e.m.f.** of the cells by **adding** their individual e.m.f.s.

$$\varepsilon_{total} = \varepsilon_1 + \varepsilon_2 + \varepsilon_3 + ...$$

This makes sense if you think about it, because each charge goes through each of the cells and so gains e.m.f. (electrical energy) from each one.

See p.84 for all the rules for parallel and series circuits.

For identical cells **in parallel** in a circuit, the **total e.m.f.** of the combination of cells is the **same size** as the e.m.f. of each of the individual cells.

$$\varepsilon_{total} = \varepsilon_1 = \varepsilon_2 = \varepsilon_3 + ...$$

This is because the current will split equally between identical cells. The charge only gains e.m.f. from the cells it travels through — so the overall e.m.f. in the circuit doesn't increase.

E.m.f. and Internal Resistance

Time for an Example **E.m.f. Calculation Question...**

Example Three identical cells with an e.m.f. of 2.0 V and an internal resistance of 0.20 Ω are connected in parallel in the circuit shown to the right. A current of 0.90 A is flowing through the circuit. Calculate the total p.d. across the cells.

First calculate the lost volts, v, for 1 cell using $v = Ir$.

Since the current flowing through the circuit is split equally between each of the three cells, the current through one cell is $I/3$. So for 1 cell: $v = I/3 \times r = 0.90/3 \times 0.20 = 0.30 \times 0.20 = 0.06$ V

Then find the terminal p.d. across 1 cell using the equation: $V = \varepsilon - v = 2 - 0.06 = 1.94$

So the total p.d. across the cells combined = 1.94 = **1.9 V (to 2 s.f.)**

Investigate **Internal Resistance** and **E.m.f.** With This **Circuit**

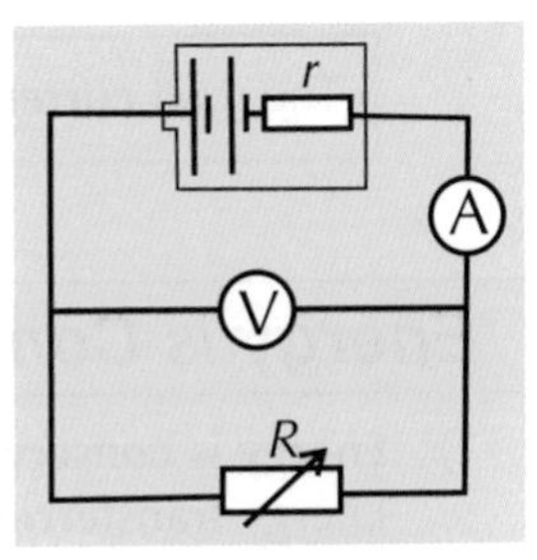

1) **Vary** the **current** in the circuit by changing the value of the **load resistance** (***R***) using the variable resistor. **Measure** the **p.d.** (***V***) for several different values of **current** (***I***).
2) Record your data for V and I in a table, and **plot the results** in a graph of V against I.

To find the **e.m.f.** and **internal resistance** of the cell, start with the equation: $V = \varepsilon - Ir$

1) Rearrange to give $V = -rI + \varepsilon$
2) Since ε and r are constants, that's just the equation of a **straight line**:

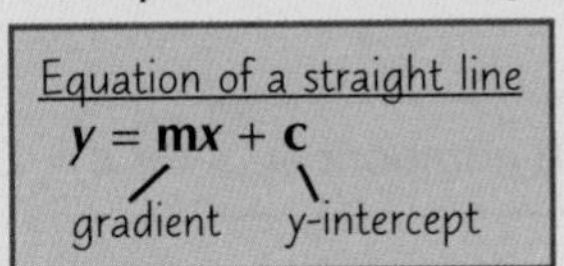

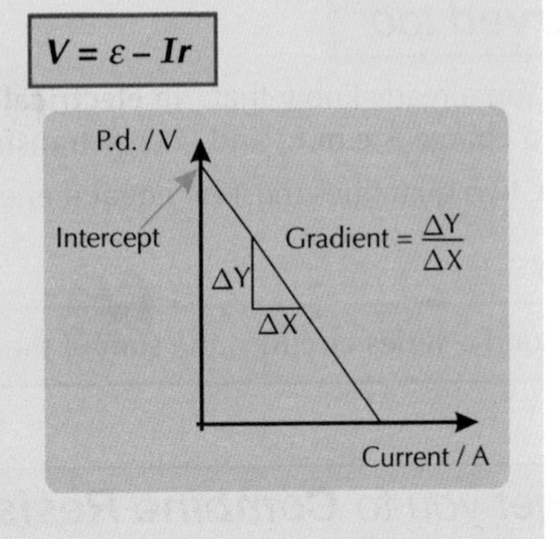

3) So the intercept on the vertical axis is ε.
4) And the gradient is $-r$.

Geoff didn't quite calculate the gradient correctly.

An **easier** way to **measure** the **e.m.f.** of a **power source** is by connecting a high-resistance **voltmeter** across its **terminals**. But, a **small current flows** through the **voltmeter**, so there must be some **lost volts** — this means you measure a value **very slightly less** than the **e.m.f.** (Although in practice the difference isn't usually significant.)

Practice Questions

Q1 What causes internal resistance? Write down the equation linking e.m.f. and energy transferred.

Q2 What is the difference between e.m.f. and terminal p.d.?

Q3 Write the equation used to calculate the terminal p.d. of a power supply.

Exam Questions

Q1 A battery with an internal resistance of 0.80 Ω and an e.m.f. of 24 V powers a dentist's drill with resistance 4.0 Ω.

a) Calculate the current in the circuit when the drill is connected to the power supply. [2 marks]

b) Calculate the potential difference wasted overcoming the internal resistance. [1 mark]

Q2 A bulb of resistance R is powered by two cells connected in series each with internal resistance r and e.m.f. ε. Which expression represents the current flowing through each cell? [1 mark]

A $\frac{\varepsilon}{R+r}$ B $\frac{\varepsilon}{2(R+2r)}$ C $\frac{2\varepsilon}{R+2r}$ D $\frac{\varepsilon}{R+2r}$

Overcome your internal resistance for revision...

Make sure you know all your e.m.f. and internal resistance equations, they're an exam fave. A good way to get them learnt is to keep trying to get from one equation to another... pretty dull, but it definitely helps.

Conservation of Energy and Charge

There are some things in Physics that are so fundamental that you just have to accept them. Like the fact that there's loads of Maths in it. And that energy is conserved. And that Physicists get more homework than everyone else.

Charge Doesn't 'Leak Away' Anywhere — it's Conserved

1) As **charge flows** through a circuit, it **doesn't** get **used up** or **lost**.
2) This means that whatever **charge flows into** a junction will **flow out** again.
3) Since **current** is **rate of flow of charge**, it follows that whatever **current flows into** a junction is the same as the current **flowing out** of it.

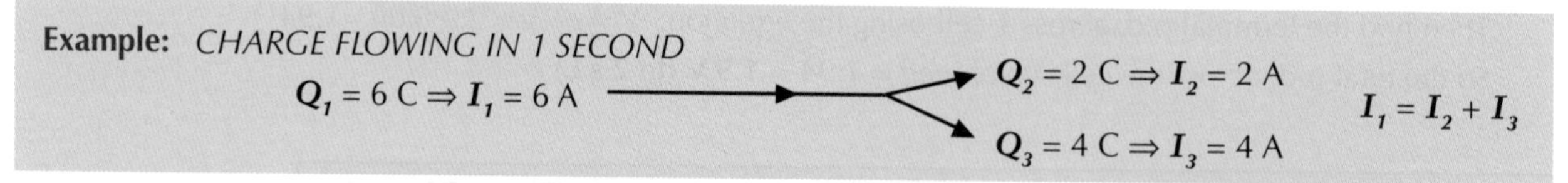

Kirchhoff's first law says:

> The total **current entering a junction** = the total **current leaving it.**

Energy conservation is vital.

Energy is Conserved too

1) **Energy is conserved**. You already know that. In **electrical circuits**, **energy** is **transferred round** the circuit. Energy **transferred to** a charge is **e.m.f.**, and energy **transferred from** a charge is **potential difference**.
2) In a **closed loop**, these two quantities must be **equal** if energy is conserved (which it is).

Kirchhoff's second law says:

> The **total e.m.f.** around a **series circuit** = the **sum** of the **p.d.s** across each component.

(or $\varepsilon = \Sigma IR$ in symbols)

Exam Questions get you to Combine Resistors in Series and Parallel

A **typical exam question** will give you a **circuit** with bits of information missing, leaving you to fill in the gaps. Not the most fun... but on the plus side you get to ignore any internal resistance stuff (unless the question tells you otherwise)... hurrah. You need to remember the **following rules**:

Series Circuits:

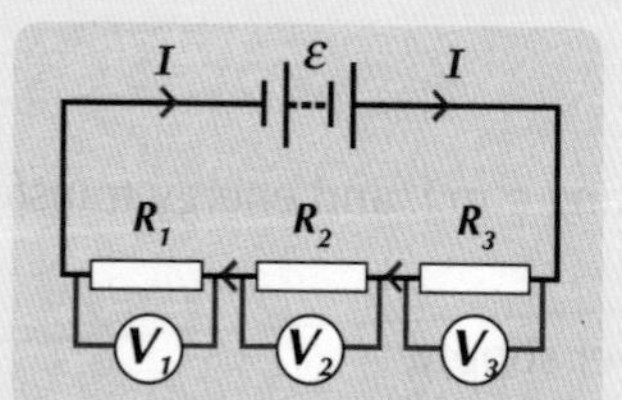

1) **same current** at **all points** of the circuit (since there are no junctions)
2) **e.m.f. split** between **components** (by Kirchhoff's 2nd law), so: $\varepsilon = V_1 + V_2 + V_3$
3) $V = IR$, so if I is constant: $IR_{total} = IR_1 + IR_2 + IR_3$
4) cancelling the Is gives:

$$R_{total} = R_1 + R_2 + R_3$$

Parallel Circuits:

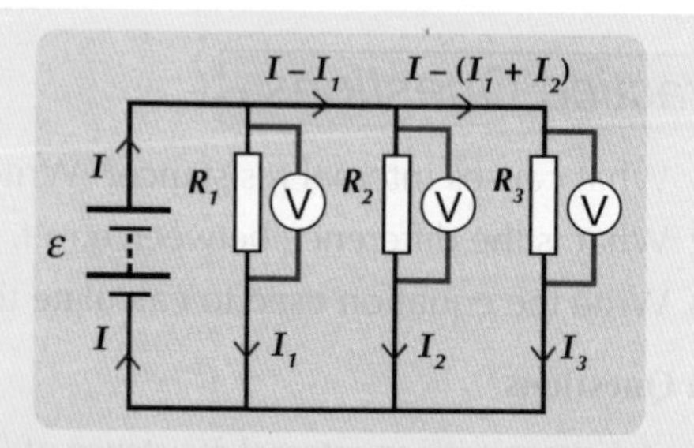

1) **current** is **split** at each **junction**, so: $I = I_1 + I_2 + I_3$
2) **same p.d.** across **all components** (three separate loops — within each loop the e.m.f. equals sum of individual p.d.s)
3) so, $V/R_{total} = V/R_1 + V/R_2 + V/R_3$
4) cancelling the Vs gives:

$$\frac{1}{R_{total}} = \frac{1}{R_1} + \frac{1}{R_2} + \frac{1}{R_3}$$

$\varepsilon = V$ in this case, as we're ignoring internal resistance.

...and there's an example on the next page to make sure you know what to do with all that...

Conservation of Energy and Charge

Worked Exam Question

Example:

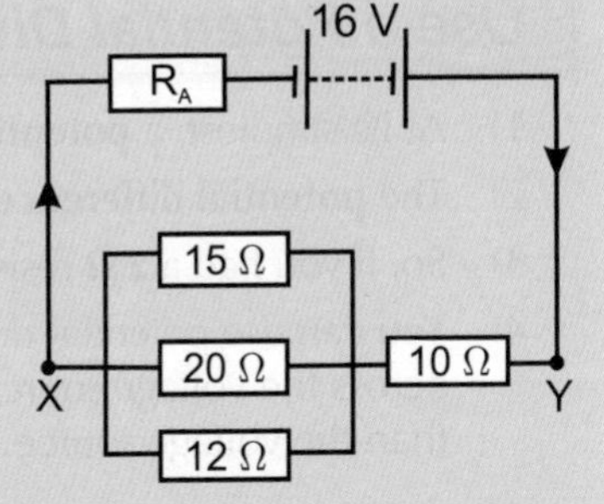

A battery of e.m.f. 16 V and negligible internal resistance is connected in a circuit as shown:

a) Show that the group of resistors between X and Y could be replaced by a single resistor of resistance 15 Ω.

You can find the combined resistance of the 15 Ω, 20 Ω and 12 Ω resistors using:
$1/R = 1/R_1 + 1/R_2 + 1/R_3 = 1/15 + 1/20 + 1/12 = 1/5 \quad \Rightarrow R = 5\ \Omega$
So overall resistance between X and Y can be found by $R = R_1 + R_2 = 5 + 10 =$ **15 Ω**

b) If $R_A = 20\ \Omega$:
i) calculate the potential difference across R_A,

Careful — there are a few steps here. You need the p.d. across R_A, but you don't know the current through it. So start there: total resistance in circuit = 20 + 15 = 35 Ω,
so current through R_A can be found using $I = V_{total}/R_{total}$: $I = 16/35$ A
then you can use $V = IR_A$ to find the p.d. across R_A: $V = 16/35 \times 20 =$ **9 V (to 1 s.f.)**

ii) calculate the current in the 15 Ω resistor.

You know the current flowing into the group of three resistors and out of it, but not through the individual branches. But you know that their combined resistance is 5 Ω (from part a) so you can work out the p.d. across the group:
$V = IR = 16/35 \times 5 = 16/7$ V
The p.d. across the whole group is the same as the p.d. across each individual resistor, so you can use this to find the current through the 15 Ω resistor:
$I = V/R = (16/7) / 15 =$ **0.15 A (to 2 s.f.)**

Practice Questions

Q1 State the formulas used to combine resistors in series and in parallel.

Q2 Find the current through and potential difference across each of two 5 Ω resistors when they are placed in a circuit containing a 5 V battery, and are wired: a) in series, b) in parallel.

Exam Question

Q1 For the circuit on the right:

a) Calculate the total effective resistance of the three resistors in this combination. [2 marks]

b) Calculate the main current, I_3. [1 mark]

c) Calculate the potential difference across the 4.0 Ω resistor. [1 mark]

d) Calculate the potential difference across the parallel pair of resistors. [1 mark]

e) Using your answer from part d), calculate the currents I_1 and I_2. [2 marks]

12 V, I_3, I_2, 6.0 Ω, 4.0 Ω, I_1, 3.0 Ω

This is a very purple page — needs a bit of yellow I think...

$V = IR$ is the formula you'll use most often in these questions. Make sure you know whether you're using it on the overall circuit, or just one specific component. It's amazingly easy to get muddled up — you've been warned.

The Potential Divider

I remember the days when potential dividers were pretty much the hardest thing they could throw at you. Then along came A level Physics. Hey ho. Anyway, in context this doesn't seem too hard now, so get stuck in.

Use a **Potential Divider** to get a **Fraction** of a **Source Voltage**

1) At its simplest, a **potential divider** is a circuit with a **voltage source** and a couple of **resistors** in series.
2) The **potential difference** across the voltage source (e.g. a battery) is **split** in the **ratio** of the **resistances** (p.84).
3) So, if you had a **2 Ω** resistor and a **3 Ω** resistor, you'd get **2/5** of the p.d. across the **2 Ω** resistor and **3/5** across the **3 Ω**.
4) You can use potential dividers to supply a potential difference, V_{out}, between **zero** and the potential difference across the voltage source. This can be useful, e.g. if you need a **varying** p.d. supply or one that is at a **lower p.d.** than the voltage source.

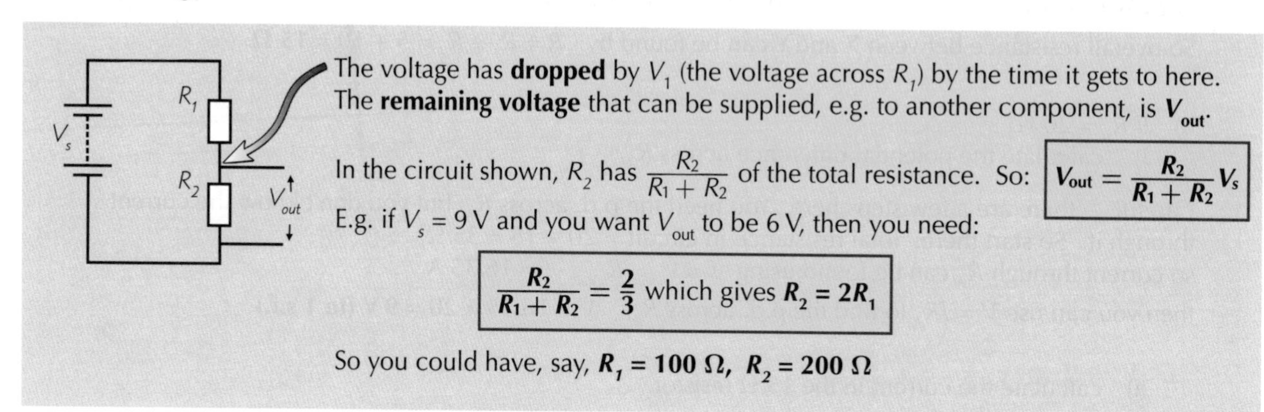

The voltage has **dropped** by V_1 (the voltage across R_1) by the time it gets to here. The **remaining voltage** that can be supplied, e.g. to another component, is V_{out}.

In the circuit shown, R_2 has $\frac{R_2}{R_1 + R_2}$ of the total resistance. So: $V_{out} = \frac{R_2}{R_1 + R_2} V_s$

E.g. if V_s = 9 V and you want V_{out} to be 6 V, then you need:

$$\frac{R_2}{R_1 + R_2} = \frac{2}{3} \text{ which gives } R_2 = 2R_1$$

So you could have, say, R_1 **= 100 Ω**, R_2 **= 200 Ω**

5) This circuit is mainly used for **calibrating voltmeters**, which have a **very high resistance**.
6) If you put something with a **relatively low resistance** across R_2 though, you start to run into **problems**. You've **effectively** got **two resistors** in **parallel**, which will **always** have a **total** resistance **less** than R_2. That means that V_{out} will be **less** than you've calculated, and will depend on what's connected across R_2. Hrrumph.

Use a **Variable Resistor** to Vary the **Voltage**

If you replace R_1 with a **variable resistor**, you can change V_{out}. When $R_1 = 0$, $V_{out} = V_s$. As you increase R_1, V_{out} gets smaller.

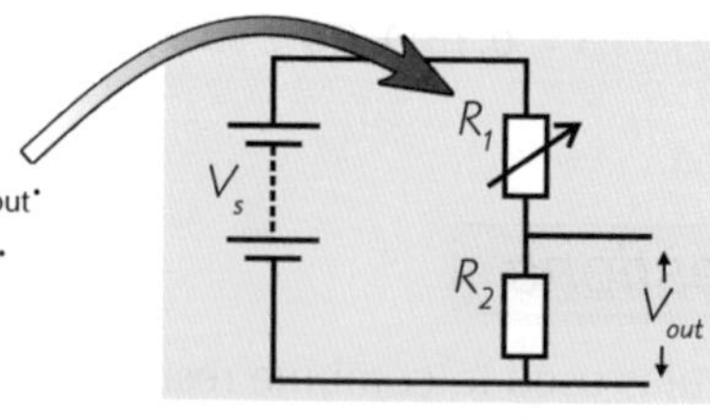

Add an **LDR** or **Thermistor** for a **Light** or **Temperature Sensor**

1) A **light-dependent resistor** (LDR) has a very **high resistance** in the **dark**, but a **lower resistance** in the **light**.

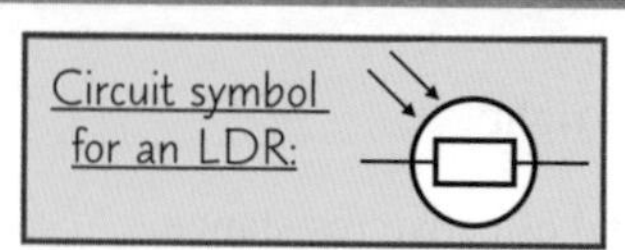

2) An **NTC thermistor** has a **high resistance** at **low temperatures**, but a much **lower resistance** at **high temperatures** (it varies in the opposite way to a normal resistor, only much more so).
3) Either of these can be used as one of the **resistors** in a **potential divider**, giving an **output voltage** that **varies** with the **light level** or **temperature**.

The diagram shows a **sensor** used to detect **light levels**. When light shines on the LDR its **resistance decreases**, so V_{out} increases.

You can include LDRs and thermistors in circuits that control **switches**, e.g. to turn on a light or a heating system.

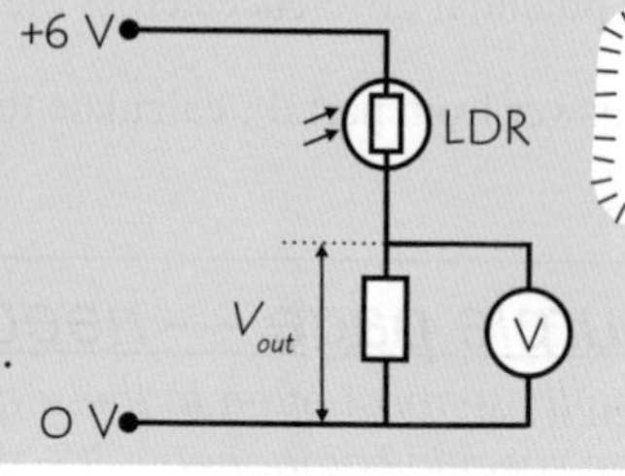

If you replace the LDR with a thermistor, V_{out} will increase with temperature.

The Potential Divider

A Potentiometer uses a Variable Resistor to give a Variable Voltage

1) A **potentiometer** has a variable resistor replacing R_1 and R_2 of the potential divider, but it uses the **same idea** (it's even sometimes **called** a potential divider just to confuse things).
2) You move a **slider** or turn a knob to **adjust** the **relative sizes** of R_1 and R_2. That way you can vary V_{out} from **0 V** up to the source voltage.
3) This is dead handy when you want to be able to **change** a **voltage continuously**, like in the **volume control** of a stereo.

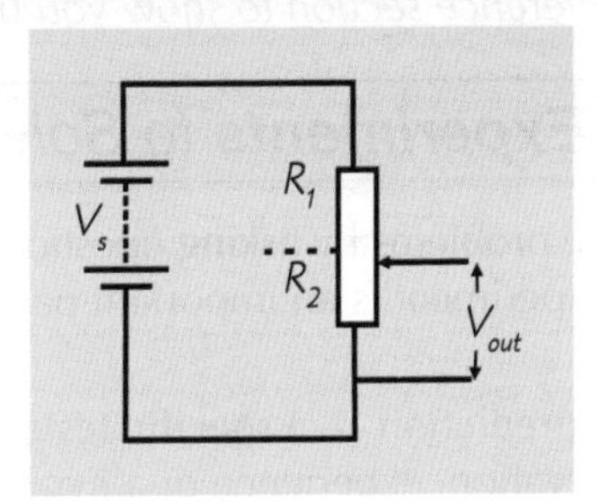

Example: Here, V_S is replaced by the input signal (e.g. from a CD player) and V_{out} is the output to the amplifier and loudspeaker.

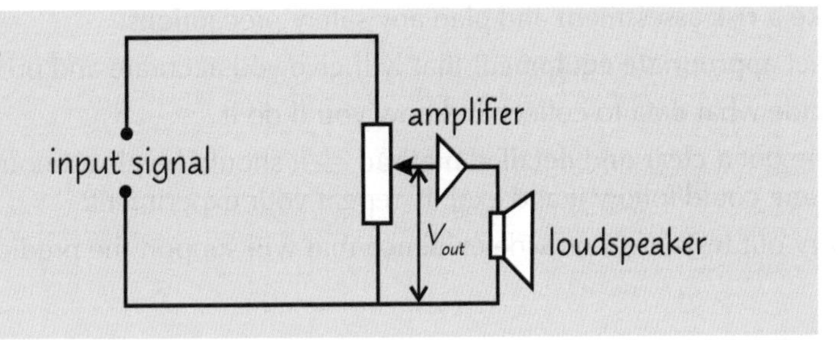

Practice Questions

Q1 Look at the light sensor circuit on page 86.
How could you change the circuit so that it could be used to detect temperature changes?

Q2 The LDR in the circuit on page 86 has a resistance of 300 Ω when in light conditions, and 900 Ω in dark conditions. The fixed resistor has a value of 100 Ω. Show that V_{out} (light) = 1.5 V and V_{out} (dark) = 0.6 V.

Exam Questions

Q1 In the circuit on the right, all the resistors have the same value.
Calculate the p.d. between:

a) A and B. [1 mark]

b) A and C. [1 mark]

c) B and C. [1 mark]

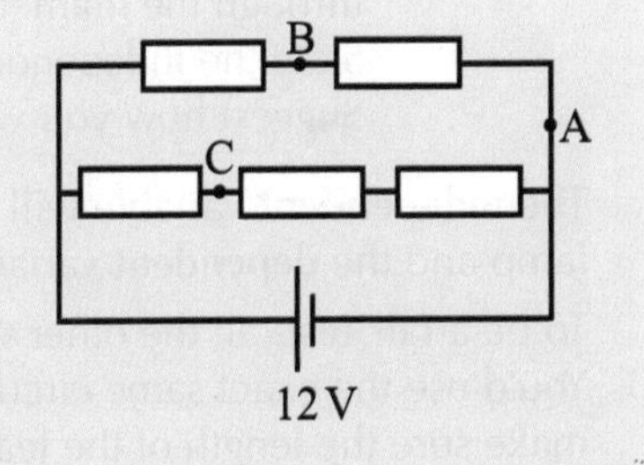

Q2 Look at the circuit on the right. All the resistances are given to 2 significant figures.

a) Calculate the p.d. between A and B as shown by a high resistance voltmeter placed between the two points. [1 mark]

b) A 40.0 Ω resistor is now placed between points A and B. Calculate the p.d. across AB and the current flowing through the 40.0 Ω resistor. [4 marks]

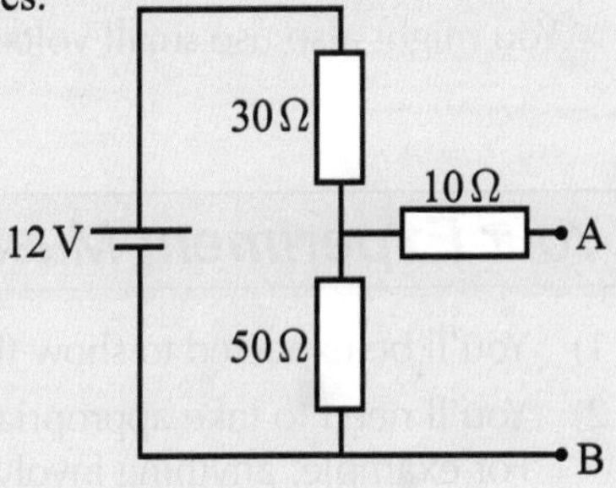

OI...YOU... [bang bang bang]... turn that potentiometer down...

You'll probably have to use a potentiometer in every experiment you do with electricity from now on in, so you'd better get used to them. I can't stand the things myself, but then lab and me don't mix — far too technical.

Experiment Design

Science is all about getting good evidence to test your theories... so you need to be able to spot a badly designed experiment or study a mile off, and be able to interpret the results of an experiment or study properly. Here's a quick reference section to show you how to go about designing experiments and doing data-style questions.

Planning **Experiments** to Solve **Problems**

Scientists solve problems by **asking** questions, **suggesting** answers and then **testing** them to see if they're correct. Planning an experiment is an important part of this process to help get accurate and precise results (see p. 93).

1) Make a **prediction** — a **specific testable statement** about what will happen in the experiment, based on observation, experience or a **hypothesis** (a **suggested explanation** for a fact or observation).
2) Think about the aims of the experiment and identify the **independent**, **dependent** and other **variables**.
3) Make a **risk assessment** and plan any safety precautions.
4) Select **appropriate equipment** that will give you accurate and precise results.
5) Decide what **data** to collect and how you'll do it.
6) Write out a **clear** and **detailed method** — it should be clear enough that **anyone** could follow it and exactly repeat your experiment.
7) Carry out **tests** — to provide **evidence** that will support the prediction or refute it.

Make Sure Your Experiment is a **Fair Test**

It's important to **control** the **variables** (any quantity that can change) in an experiment. Keeping all variables **constant** apart from the independent and dependent variables, means that the experiment is a **fair test**. This means you can be more confident that any effects you see are **caused** by changing the independent variable. The variables that are kept constant (or at least monitored) in an experiment are called **control variables**.

Independent variable
The thing that you **change** in an experiment.

Dependent variable
The thing that you **measure** in an experiment.

Example: The circuit on the right is used to investigate how the current through the filament lamp varies with the voltage across it. State the independent and dependent variables in this experiment. Suggest how you would make this experiment a fair test.

Power supply. V_s

Ammeter

Potentiometer (p. 87) to vary the supplied voltage

Filament lamp

Voltmeter

The **independent** variable will be the **voltage** across the lamp and the **dependent** variable will be the **current**.

To be a **fair test**, all the other variables must be kept the **same**. You'd use the exact same circuit throughout the experiment to make sure the length of the leads, the filament bulb and the resistance of the rest of the circuit all remained the same.

You might also use small voltages (and currents) to stop the circuit wires heating up during the experiment.

Your **Experiment** Must Be **Safe** and **Ethical**

1) You'll be expected to show that you can identify any **risks** and **hazards** in an experiment.
2) You'll need to take appropriate **safety measures** depending on the experiment. For example, anything involving **lasers** will usually need special laser **goggles** and to work with **radioactive substances** you'll probably need to wear **gloves**.
3) You need to make sure you're working **ethically** too — you've got to look after the **welfare** of any people or animals in an experiment to make sure they don't become **ill**, **stressed** or **harmed** in any way.
4) You also need to make sure you're treating the **environment ethically** too, e.g. making sure to not destroy habitats when doing **outdoor** experiments.

Monty was stressed out by the velocity experiment before it even began.

Experiment Design

Nothing is Certain

1) **Every** numerical measurement you take has an **experimental uncertainty**. The smallest uncertainty you can have in a measurement is ± **half** of one division on the measuring instrument used. E.g. using a thermometer with a scale where each division represents 2 °C, a measurement of 30 °C will at **best** be measured to be 30 **± 1 °C**. And that's without taking into account any other errors that might be in your measurement.
2) The ± sign gives you the **range** in which the **true** length (the one you'd really like to know) probably lies — 30 ± 0.5 cm tells you the true length is very likely to lie in the range of 29.5 to 30.5 cm.
3) There are **two types** of **error** that cause experimental uncertainty:

Random errors

1) Random errors cause readings to be **spread** about the true value due to the results varying in an **unpredictable** way. They affect **precision** (see p.93).
2) They can just be down to **noise**, or measuring a **random process** such as nuclear radiation emission. No matter how hard you try, you **can't** correct them.
3) You get random error in **any** measurement. If you measured the length of a wire 20 times, the chances are you'd get a **slightly different** value each time, e.g. due to your head being in a slightly different position when reading the scale.
4) It could be that you just can't keep controlled variables **exactly** the same throughout the experiment.

Systematic errors

1) Systematic errors cause each reading to be different to the true value by the **same amount** i.e. they **shift** all of your results. They affect the **accuracy** of your results (see p.93).
2) Systematic errors are caused by the **environment**, the **apparatus** you're using, or your experimental method, e.g. using an inaccurate clock.
3) The problem is often that you **don't know they're there**. You've got to spot them first to have any chance of correcting for them.
4) If you **suspect** a systematic error, you should **repeat** the experiment with a different **technique** or **apparatus** and compare the results.

There are Loads of Ways You Can Reduce Uncertainties

1) One of the easiest things you can do is **repeat** each measurement **several times**. The **more repeats** you do, and the more **similar** the results of each repeat are, the more precise the data.
2) By taking the **mean** of repeated measurements, you will reduce the **random error** in the result. You calculate a mean by adding up all of the measurements and dividing by the total number of measurements. The **more** measurements you average over, the **less random error** you're likely to have.
3) The **smaller the uncertainty** in a result or measurement, the **smaller the range** of possible values that result could have and the more **precise** your data can be. E.g. two students each measure a length of wire three times. Student A measures the wire to be 30 cm ± 1 cm each time. Student B measures the wire to be 29 cm ± 0.5 cm each time. The **range** that student A's values could take is **larger** than student B's, so student B's data is more **precise**.
4) You should check your data for any **anomalous** results — any results that are **so different** from the **rest of the data** they cannot be explained as variations caused by random uncertainties. For example, a measurement is ten times smaller than all of your other data values. You should not include anomalous results when you take averages.
5) You can also cut down the **uncertainty** in your measurements by using the most **appropriate** equipment. E.g. a micrometer scale has **smaller intervals** than a millimetre ruler — so by measuring a wire's diameter with a micrometer instead of the ruler, you instantly cut down the **random error** in your experiment.
6) **Computers** and **data loggers** can often be used to measure smaller intervals than you can measure by hand and reduce random errors, e.g. timing an object's fall using a light gate rather than a stop watch. You also get rid of any **human error** that might creep in while taking the measurements.
7) There's a limit to how much you can reduce the random uncertainties in your measurements, as all measuring equipment has a **resolution** — the smallest change in what's being measured that can be detected by the equipment.
8) You can **calibrate** your apparatus by measuring a **known value**. If there's a **difference** between the **measured** and **known** value, you can use this to **correct** the inaccuracy of the apparatus, and so reduce your **systematic error**. For example, to calibrate a set of **scales** you could weigh a 10.0 g mass and check that it reads 10.0 g. If these scales measure to the nearest 0.1 g, then you can only **calibrate** to within 0.05 g. Any measurements taken will have an **uncertainty** of ± 0.05 g.
9) **Calibration** can also reduce **zero errors** (caused by the apparatus **failing to read zero** when it should do, e.g. when no current is flowing through an ammeter) which can cause systematic errors.

I'm certain that I need a break...

There's a lot to take in here. Make sure you remember the different kinds of errors and how you can avoid them...

Uncertainty and Errors

Significant figures, uncertainties and error bars are all ways of saying how almost-sure you are about stuff.

Uncertainties Come in *Absolute Amounts, Fractions* and *Percentages*

Absolute uncertainty is the uncertainty of a measurement given as certain fixed quantity.
Fractional uncertainty is the uncertainty given as a **fraction** of the measurement taken.
Percentage uncertainty is the uncertainty given as a **percentage** of the measurement.

An uncertainty should also include a **level of confidence** or **probability**, to indicate how **likely** the true value is to lie in the interval. E.g. '5.0 ± 0.4 Ω at a level of confidence of 80%' means you're **80% sure** that the true value is **within** 0.4 Ω of 5.0 Ω. (Don't worry, you **don't need** to calculate the level of confidence.)

Example: The resistance of a filament lamp is given as 5.0 ± 0.4 Ω. Give the absolute, fractional and percentage uncertainties for this measurement.

1) The absolute uncertainty is given in the question — it's **0.4 Ω.**
2) To calculate fractional uncertainty, divide the uncertainty by the measurement and simplify. The fractional uncertainty is $\frac{0.4}{5.0} = \frac{4}{50} = \mathbf{\frac{2}{25}}$
3) To calculate percentage uncertainty, divide the uncertainty by the measurement and multiply by 100. The percentage uncertainty is $\frac{2}{25} \times 100 = \mathbf{8\%}$

You can **decrease** the **percentage uncertainty** in your data by taking measurements of **large** quantities. Say you take measurements with a ruler which measures to the nearest **± 0.5 mm**. The **percentage error** in measuring **10 mm** will be **± 5%**, but using the same ruler to measure a distance of **20 cm** will give a percentage error of only **± 0.25%**.

The uncertainty on a **mean** (see p.89) of repeated results is equal to **half the range** of the results.
E.g. say the repeated measurement of a current gives the results 0.5 A, 0.3 A, 0.3 A, 0.3 A and 0.4 A.
The range of these results is 0.5 – 0.3 = 0.2 A, so the uncertainty on the mean current would be **± 0.1 A.**

Sometimes You Need to *Combine Uncertainties*

When you do calculations involving values that have an uncertainty, you have to **combine** the uncertainties to get the **overall** uncertainty for your result.

Adding or *Subtracting* Data — *ADD* the *Absolute Uncertainties*

Example: A wire is stretched from 0.3 ± 0.1 cm to 0.5 ± 0.1 cm. Calculate the extension of the wire.

1) First subtract the lengths without the uncertainty values: 0.5 – 0.3 = 0.2 cm
2) Then find the total uncertainty by adding the individual absolute uncertainties: 0.1 + 0.1 = 0.2 cm
   So, the wire has been stretched **0.2 ± 0.2 cm.**

Multiplying or *Dividing* Data — *ADD* the *Percentage Uncertainties*

Example: A force of 15 ± 3% N is applied to a stationary object which has a mass of 6.0 ± 0.3 kg. Calculate the acceleration of the object and state the percentage uncertainty in this value.

1) First calculate the acceleration without uncertainty: $a = F \div m = 15 \div 6.0 = 2.5\ \text{ms}^{-2}$
2) Next, calculate the percentage uncertainty in the mass: % uncertainty in $m = \frac{0.3}{6} \times 100 = 5\%$
3) Add the percentage uncertainties in the force and mass values to find the total uncertainty in the acceleration: Total uncertainty = 3% + 5% = 8%
   So, the acceleration = **2.5 ms^{-2} ± 8%**

Raising to a Power — *MULTIPLY* the *Percentage Uncertainty* by the *Power*

Example: The radius of a circle is r = 40 cm ± 2.5%. What will the percentage uncertainty be in the area of this circle, i.e. πr^2?

The radius will be raised to the power of **2** to calculate the area.
So, the percentage uncertainty will be 2.5% × 2 = **5%**

Uncertainty and Errors

Error Bars Show the Uncertainty of Individual Points

1) Most of the time, you work out the **uncertainty** in your **final** result using the uncertainty in **each measurement** you make.
2) When you're plotting a **graph**, you show the uncertainty in **each measurement** by using **error bars** to show the **range** the point is likely to lie in.
3) In exams, you might have to **analyse data** from graphs **with** and **without** error bars — so make sure you really understand what error bars are showing.
4) The **error** of **each measurement** when measuring the extension of a material is shown by the **error bars** in the graph to the right.

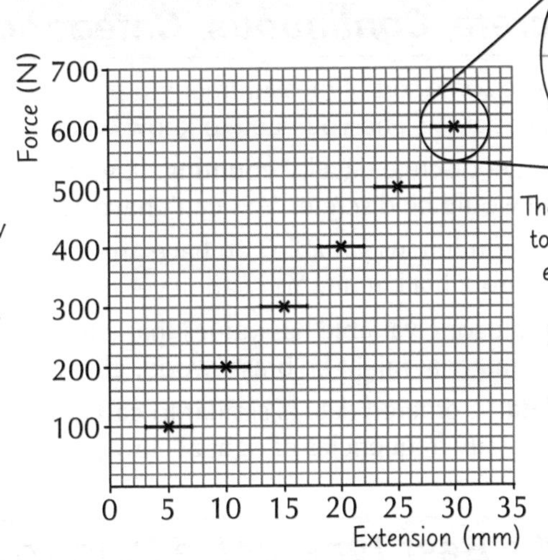

The error bars extend 2 squares to the right and to the left for each measurement, which is equivalent to 2 mm. So, the uncertainty in each measurement is ± 2 mm.

Your line of best fit (p.92) should always go through all of the error bars.

You Can Calculate the Uncertainty of Final Results from a Line of Best Fit

Normally when you draw a graph you'll want to find the **gradient** or **intercept**. For example, you can calculate k, the **spring constant** of the object being stretched, from the **gradient** of the graph on the right — here it's about 20 000 Nm^{-1}. You can find the **uncertainty** in that value by using **worst lines**:

1) Draw lines of best fit which have the **maximum** and **minimum** possible slopes for the data and which should go through all of the **error bars** (see the pink and blue lines on the right). These are the **worst lines** for your data.
2) Calculate the **worst gradient** — the gradient of the slope that is **furthest** from the gradient of the line of best fit. The blue line's gradient is about 21 000 Nm^{-1} and the pink line's gradient is about 19 000 Nm^{-1}, so you can use either here.
3) The **uncertainty** in the gradient is given by the **difference** between the **best gradient** (of the line of best fit) and the **worst gradient** — here it's 1000 Nm^{-1}. So this is the uncertainty in the value of the spring constant. For this object, the spring constant is 20 000 ± 1000 Nm^{-1} (or 20 000 Nm^{-1} ± 5%).
4) Similarly, the uncertainty in the ***y*-intercept** is just the **difference** between the **best** and **worst** intercepts (although there's no uncertainty here since the best and worst lines both go through the origin).

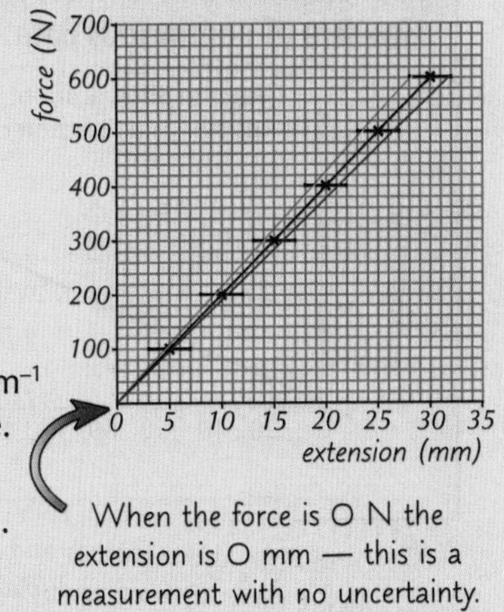

When the force is 0 N the extension is 0 mm — this is a measurement with no uncertainty.

Significant Figures Can Show Uncertainty

1) You always have to assume the **largest** amount of uncertainty in data.
2) Whether you're looking at experimental results or just doing a calculation question in an exam, you must round your results to have the **same number** of significant figures as the given data value with the **least** significant figures. Otherwise you'd be saying there is less uncertainty in your result than in the data used to calculate it.
3) If no uncertainty is given for a value, the number of **significant figures** a value has gives you an estimate of the **uncertainty**. For example, 2 N only has **1 significant figure**, so without any other information you know this value must be 2 **± 0.5 N** — if the value was less than 1.5 N it would have been rounded to 1 N (to 1 s.f.), if it was greater than 2.5 N it would have been rounded to 3 N (to 1 s.f.).

I'd give uncertainties 4 ± 2 for fun...

There's lots of maths to get your head around here, but just keep practising calculating uncertainties and you'll learn the rules in no time. Well... t = 0 ± 4 hours, sorry. Have another read and flip the book over, then scribble down the key points you can remember. Keep doing it until you can remember all the uncertainty joy without having to sneak a look.

Presenting and Evaluating Data

Once you've got results, you have to present them in a sensible way using a graph. Then it's time to evaluate them and use them to form a conclusion that is supported by your results.

Data can be **Discrete, Continuous, Categoric** or **Ordered**

Experiments always involve some sort of measurement to provide data.
There are different types of data — and you need to know what they are.

1) **Discrete data** — you get discrete data by **counting**. E.g. the number of weights added to the end of a spring would be discrete. You can't have 1.25 weights.
2) **Continuous data** — a continuous variable can have **any value** on a scale. For example, the extension of a spring or the current through a circuit. You can never measure the exact value of a continuous variable.
3) **Categoric data** — a categoric variable has values that can be **sorted** into **categories**. For example, types of material might be brass, wood, glass, steel.
4) **Ordered (ordinal) data** — ordered data is similar to **categoric**, but the categories can be put **in order**. For example, if you classify frequencies of light as 'low', 'fairly high' and 'very high' you'd have ordered data.

Graphs — Use the **Best Type** for the Data You've Got

You'll usually be expected to make a **graph** of your results. Not only are graphs **pretty**, they make your data **easier to understand** — so long as you choose the right type. No matter what the type though, make sure you always **label your axes** — including **units**. Choose a **sensible scale** for your axes and **plot points accurately** using a sharp pencil.

Line graphs are best when you have **two sets of continuous data**. For example:

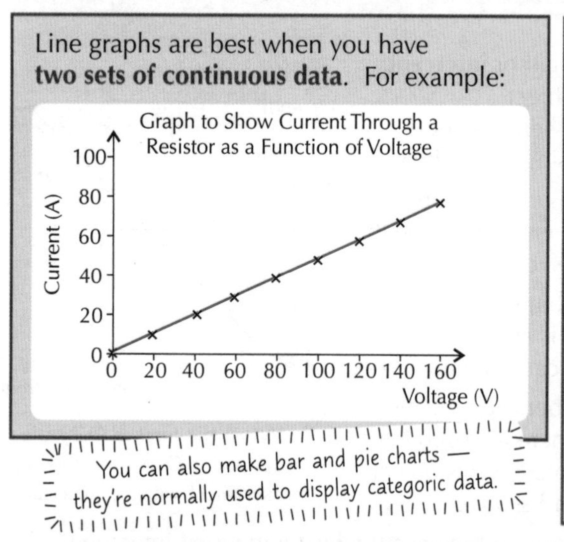

You can also make bar and pie charts — they're normally used to display categoric data.

Scatter plots are great for showing how two sets of data are related (or **correlated**). Don't try to join all the points — draw a **line of best fit** to show the **trend**.

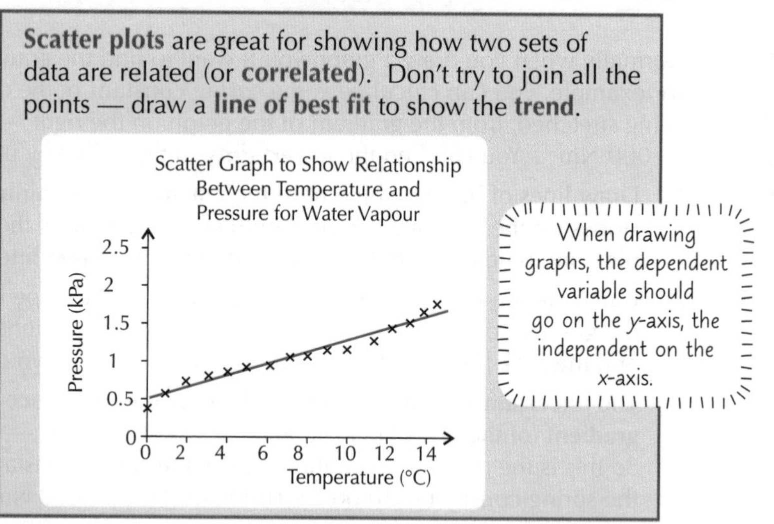

When drawing graphs, the dependent variable should go on the *y*-axis, the independent on the *x*-axis.

Correlation Shows **Trends** in Data

1) **Correlation** describes the relationship between **two variables** — usually the **independent** and **dependent** ones.
2) Data can show **positive, negative** or **no correlation.** An easy way to see correlation is to plot a **scatter graph** of your data. If you can, draw a **line of best fit** to help show the **trend.**

Positive correlation — as one variable **increases,** the other also **increases.**

Negative correlation — as one variable **increases,** the other **decreases.**

No correlation — there is **no relationship** between the variables.

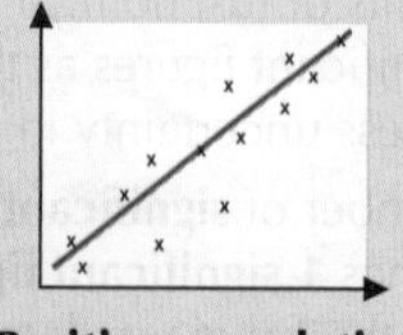

Positive correlation

Negative correlation

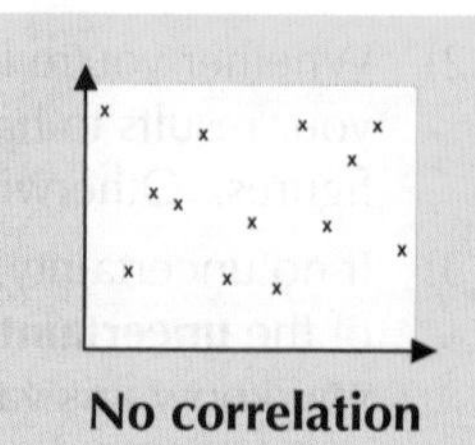

No correlation

3) If you've done a **controlled** experiment in a lab and can see **correlation** in your results, you can be fairly certain there's a **causal relationship** between the **independent** and **dependent variables.** This means that a **change** in one **causes** a change in the other.
4) But in experiments or studies **outside** the lab, you can't usually control all the variables. So even if two variables are **correlated,** the change in one may not be causing the change in the other. Both changes might be caused by a **third** variable.

Presenting and Evaluating Data

Evaluating Your Data

Once an experiment's over, you have to **explain** what the data shows.
There are some key words you need to know about (and use) when evaluating data:

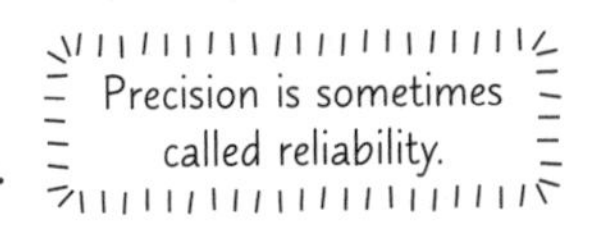

1) **Precision** — the smaller the amount of **spread** of your data from the **mean**, the more precise it is. Precision only depends on the amount of **random error** in your readings.
2) **Repeatable** — **you** can **repeat** an experiment multiple times and get the **same results**.
3) **Reproducible** — if **someone else** can recreate your experiment using different equipment or methods, and gets the **same results** you do, the results are reproducible.
4) **Valid** — the **original question** is **answered**. If you don't keep all variables apart from the one you're testing **constant**, you haven't **only** tested the variable you're investigating and so the results **aren't valid**.
5) **Accurate** — the result is really close to the **true answer**.
   You can only comment on how accurate a result is if you know the true value of the result.

There's normally loads of stuff to say when you're looking at data. Have a think about...

1) What **patterns** or trends, if any, the results show.
2) Whether the experiment managed to **answer** the question it set out to answer.
   If it did, is this a **valid** experiment and if not, why not? How **precise** was the data?
3) How close the results are to the **true value**.
4) Did the measuring instruments have enough **resolution**?
5) Any **anomalies** in the results and the possible causes of them.
6) How **large** the **uncertainties** are. If the percentage uncertainty is large, this suggests the data is not precise and a strong conclusion cannot be made.

Nora telling everyone she was 35 was a little inaccurate.

If you're asked to analyse data in the exam, look at how many marks the question is worth — the more **marks** allocated to the question in the exam, the **more detail** you have to go into.

Drawing **Conclusions** From Your **Data**

You need to make sure your conclusion is **specific** to the data you have and is **supported** by the data — don't go making any sweeping generalisations.
Your conclusion is only **valid** if it is supported by **valid data**, known as **evidence**.

Example: The stress of a material X was measured at strains of 0.002, 0.004, 0.006, 0.008 and 0.010. Each strain reading had an error of 0.001. All other variables were kept constant. A science magazine concluded from a graph of this data that material X's yield point is at a strain of 0.005. Explain whether or not you agree with this conclusion.

Their conclusion **could** be true — but the **data doesn't support this**. You can't tell **exactly** where the yield point is from the data because strain increases of 0.002 at a time were used. The stress at in-between strains wasn't measured — so all you know is that the yield point is somewhere **between** 0.004 and 0.006, as the stress drops between these values.

Stress (Nm^{-2})

0 0.2 0.4 0.6 0.8 1.0

Strain ($\times 10^{-2}$)

Also, the graph only gives information about this particular experiment. You can't conclude that the yield point would be in this range for **all experiments** — only this one. And you can't say for sure that doing the experiment at, say, a **different constant temperature** wouldn't give a different yield point.

The error in each reading is 0.001, which gives a **percentage uncertainty** of 50% for the lowest strain reading. This is very large and could mean the results are not valid, so no definite conclusions can be drawn from them.

Am I correlated? Well I suppose I'm pretty trendy...

Remember to evaluate the data thoroughly and when you make your conclusions, always back up your points using those evaluations. Use the checklist above to give you an idea of what to say. Suggest ways in which the experiment could be improved — there's some overlap here with how to improve uncertainties, so practising this is extra useful.

Answers

Section 1 — Particles

Page 3 — Atomic Structure

1 Inside every atom there is a nucleus which contains protons and neutrons ***[1 mark]***. Orbiting this core are the electrons ***[1 mark]***.

2 Proton number is 8, so there are 8 protons and electrons (it's neutral) ***[1 mark]***. The nucleon number is 16. This is the total number of protons and neutrons. Subtract the 8 protons and that leaves 8 neutrons ***[1 mark]***.

3 a) Atoms with the same number of protons but different numbers of neutrons are called isotopes ***[1 mark]***.

b) Any two from: e.g. They have the same chemical properties. / Their nuclei have different stabilities. / They have different physical properties. ***[2 marks available — 1 mark for each correct answer.]***

4 A $^{4}_{2}$He nucleus has 2 protons and 2 neutrons (but no electrons) ***[1 mark]***.
Charge = $2 \times 1.60 \times 10^{-19} = 3.20 \times 10^{-19}$ C ***[1 mark]***
Mass = $4 \times 1.67 \times 10^{-27} = 6.68 \times 10^{-27}$ kg ***[1 mark]***

$$\text{Specific charge} = \frac{\text{charge}}{\text{mass}} = \frac{3.20 \times 10^{-19}}{6.68 \times 10^{-27}} = 4.790... \times 10^{7}$$
$$= \mathbf{4.79 \times 10^{7}\ C\ kg^{-1}}\ \textbf{(to 3 s.f.)}$$ ***[1 mark]***

The actual mass of a helium nucleus is slightly less than this due to energy being released when the nucleus was made. However, you don't need to worry about this at this level.

Page 5 — Stable and Unstable Nuclei

1 a) The strong nuclear force must be repulsive at very small nucleon separations to prevent the nucleus being crushed to a point ***[1 mark]***.

b) The protons repel each other with an electrostatic force and attract each other with the nuclear strong force. The strong force is not large enough to overcome this repulsion ***[1 mark]***. When two neutrons are added to the nucleus, they attract each other and the protons via the strong force. The strong force is now able to balance out the force of repulsion between the protons ***[1 mark]***.

2 a) $^{226}_{88}\text{Ra} \rightarrow\ ^{222}_{86}\text{Rn} +\ ^{4}_{2}\alpha$ ***[3 marks available — 1 mark for the alpha particle, 1 mark for the proton number of radon and 1 mark for the nucleon number of radon.]***

b) $^{40}_{19}\text{K} \rightarrow\ ^{40}_{20}\text{Ca} +\ ^{0}_{-1}\beta + \bar{\nu}_e$ ***[4 marks available — 1 mark for the beta particle, 1 mark each for the proton number and nucleon number of calcium, 1 mark for the antineutrino.]***

Page 7 — Particles and Antiparticles

1 $e^{+} + e^{-} \rightarrow \gamma + \gamma$ ***[1 mark]*** This is called annihilation ***[1 mark]***

2 Energy and mass are equivalent ***[1 mark]***. When two particles collide, there is a lot of energy at the point of impact. This energy is converted to mass ***[1 mark]***.

3 The creation of a particle of matter also requires the creation of its antiparticle. In this case no antineutron has been produced ***[1 mark]***.

4 Total energy = $2 \times 9.84 \times 10^{-14} = 1.968 \times 10^{-13}$ J ***[1 mark]***

$$E = hf, \text{ so } f = \frac{E}{h} = \frac{1.968 \times 10^{-13}}{6.63 \times 10^{-34}} = 2.968... \times 10^{20}$$
$$= \mathbf{2.97 \times 10^{20}\ Hz}\ \textbf{(to 3 s.f.)}$$ ***[1 mark]***

Page 9 — Forces and Exchange Particles

1 The electrostatic force is due to the exchange of virtual photons that only exist for a very short time ***[1 mark]***. The force is due to the momentum transferred to or gained from the photons as they are emitted or absorbed by a proton ***[1 mark]***.

2

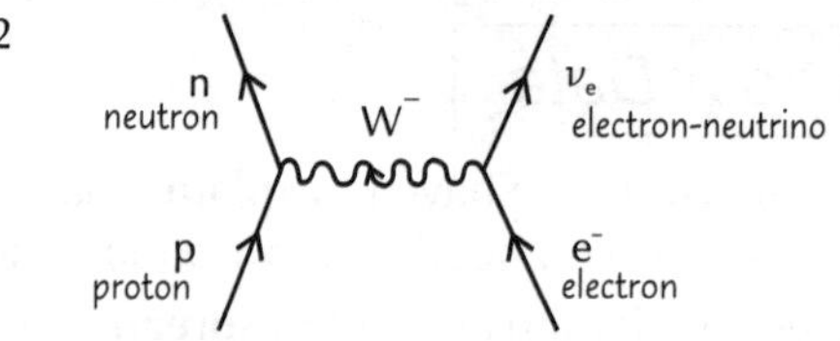

[1 mark for showing a W⁻ boson being exchanged from the electron to the proton, 1 mark for showing a neutron is produced.]
This is a weak interaction ***[1 mark]***.
Don't forget to put the arrows on your diagram.
Remember that the weak interaction uses W bosons, and is the only force that can change protons into neutrons or neutrons into protons.

Page 12 — Classification of Particles

1 Proton, electron and electron-antineutrino ***[1 mark]***. The electron and the electron-antineutrino are leptons ***[1 mark]***. Leptons are not affected by the strong interaction, so the decay can't be due to the strong interaction ***[1 mark]***.
Remember that it's really just the same as beta decay.
Some books might leave out the antineutrino, so don't be misled.

2 Mesons are hadrons but the muon is a lepton ***[1 mark]***. The muon is a fundamental particle but mesons are not. Mesons are built up from simpler particles ***[1 mark]***. Mesons interact via the strong interaction but the muon does not ***[1 mark]***.
You need to classify the muon correctly first and then say why it's different from a meson because of what it's like and what it does.

3 The correct answer is **B** ***[1 mark]***.
Protons and neutrons are both baryons, which are a type of hadron. Electrons are leptons.

Page 15 — Quarks

1 a) $\pi^{-} = d\bar{u}$ ***[1 mark]***

b) Charge of down quark = –1/3 unit.
Charge of anti-up quark = –2/3 unit.
Total charge = –1 unit ***[1 mark]***

2 The weak interaction converts a down quark into an up quark plus an electron and an electron-antineutrino ***[1 mark]***.
The neutron (udd) becomes a proton (uud) ***[1 mark]***.
The lepton number L_e is conserved in this reaction.

3 The Baryon number changes from 2 to 1, so baryon number is not conserved ***[1 mark]***. The strangeness changes from 0 to 1, so strangeness is not conserved ***[1 mark]***.

Section 2 — Electromagnetic Radiation and Quantum Phenomena

Page 17 — The Photoelectric Effect

1 The plate becomes positively charged ***[1 mark]***.
Negative electrons in the metal absorb energy from the UV light and leave the surface ***[1 mark]***.
There's one mark for saying what happens, and a second mark for saying why.

2 An electron needs to gain a certain amount of energy (the work function) before it can leave the surface of the metal ***[1 mark]***. If the energy carried by each photon is less than this work function, no electrons will be emitted ***[1 mark]***.

Page 19 — Energy Levels and Photon Emission

1 a) i) E (eV) = V = 12.1 eV ***[1 mark]***

ii) E (J) = E (eV) × (1.60×10^{-19}) = 12.1 × (1.60×10^{-19})
= $\mathbf{1.94 \times 10^{-18}}$ **J (to 3 s.f.)** ***[1 mark]***

b) i) The movement of an electron from a lower energy level to a higher energy level by absorbing energy ***[1 mark]***.

ii) –13.6 + 12.1 = –1.5 eV. This corresponds to n = 3 ***[1 mark]***.

Answers

iii) n = 3 → n = 2: –1.5 – (–3.4) = **1.9 eV** *[1 mark]*
n = 2 → n = 1: –3.4 – (–13.6) = 10.2 eV = **10 eV (to 2 s.f.)** *[1 mark]*
n = 3 → n = 1: –1.5 – (–13.6) = 12.1 = **12 eV (to 2 s.f.)** *[1 mark]*

Page 21 — Wave-Particle Duality

1 a) Electromagnetic radiation can show characteristics of both a particle and a wave. *[1 mark]*

b) $\lambda = \frac{h}{mv}$ so $mv = \frac{h}{\lambda} = \frac{6.63\times10^{-34}}{590\times10^{-9}}$ *[1 mark]*
$= \mathbf{1.1\times10^{-27}\ kgms^{-1}}$ **(to 2 s.f.)** *[1 mark]*

2 a) $\lambda = \frac{h}{mv} = \frac{6.63\times10^{-34}}{9.11\times10^{-31}\times3.50\times10^{6}}$ *[1 mark]*
$= 2.079...\times10^{-10} = \mathbf{2.08\times10^{-10}\ m}$ **(to 3 s.f.)** *[1 mark]*

b) Either $v = \frac{h}{m\lambda} = \frac{6.63\times10^{-34}}{1.67\times10^{-27}\times2.079...\times10^{-10}}$ *[1 mark]*
$= 1909.28... = \mathbf{1910\ ms^{-1}}$ **(to 3 s.f.)** *[1 mark]*
Or momentum of protons = momentum of electrons
so $m_p \times v_p = m_e \times v_e$
$v_p = v_e \times \frac{m_e}{m_p} = 3.50\times10^{6}\times\frac{9.11\times10^{-31}}{1.67\times10^{-27}}$ *[1 mark]*
$= 1909.28... = \mathbf{1910\ ms^{-1}}$ **(to 3 s.f.)** *[1 mark]*

c) The proton has a larger mass, so it will have a smaller speed, since the two have the same kinetic energy *[1 mark]*. Kinetic energy is proportional to the square of the speed, while momentum is proportional to the speed, so they will have different momenta *[1 mark]*. Wavelength depends on the momentum, so the wavelengths will be different *[1 mark]*.
This is a really hard question. If you didn't get it right, make sure you understand the answer fully. Do the algebra if it helps.

Section 3 — Waves

Page 23 — Progressive Waves

1 a) Use $c = \lambda f$ and $f = 1/T$, so $c = \lambda / T$, giving $\lambda = cT$
$\lambda = 3.0\ ms^{-1} \times 6.0\ s$ *[1 mark]* = **18 m** *[1 mark]*
The vertical movement of the buoy is irrelevant to this part of the question.

b) The trough-to-peak distance is twice the amplitude, so the amplitude is **0.6 m** *[1 mark]*

2 $c = \lambda f$, so $f = c/\lambda = (3.00\times10^{8}) \div (7.1\times10^{-7})$
$= 4.225...\times10^{14} = \mathbf{4.2\times10^{14}\ Hz}$ **(to 2 s.f.)** *[1 mark]*

3 **B** *[1 mark]*

Page 25 — Longitudinal and Transverse Waves

1 a) The reflected light has been partially polarised *[1 mark]*. Only transverse waves can be polarised *[1 mark]*.

b) Polaroid material only transmits vibrations in one direction *[1 mark]*. Reflected light mostly vibrates in one direction, so Polaroid sunglasses filter out that direction, reducing glare *[1 mark]*.

2 Sound is a longitudinal wave *[1 mark]*. The vibrations are in the same direction as the energy transfer, so it cannot be polarised *[1 mark]*.

Page 27 — Superposition and Coherence

1 a) The frequencies and wavelengths of the two sources are equal *[1 mark]* and the phase difference is constant *[1 mark]*.

b) Interference will only be noticeable if the amplitudes of the two waves are approximately equal *[1 mark]*.

2 **B** *[1 mark]*
Remember, displacement and velocity are vector quantities — when two points on a wave are exactly out of phase, the phase difference is 180° and the velocity and displacement of the points are equal in size, but opposite in direction.

Page 29 — Stationary Waves

1 a) The length of the string for a stationary wave at the fundamental frequency is half the wavelength of the wave *[1 mark]*, so $\lambda = 2 \times 1.2 = \mathbf{2.4\ m}$ *[1 mark]*.

b) $f_{new} = \frac{1}{2l}\sqrt{\frac{2T}{\mu}} = \sqrt{2}\times\left(\frac{1}{2l}\sqrt{\frac{T}{\mu}}\right) = \sqrt{2}\times f_{original}$ *[1 mark]*
$f_{new} = \sqrt{2}\times10$ *[1 mark]* = 14.142...
= **14 Hz (to 2 s.f.)** *[1 mark]*

c) When the string forms a standing wave, its amplitude varies from a maximum at the antinodes to zero at the nodes *[1 mark]*. In a progressive wave all the points vibrate at the same amplitude *[1 mark]*.

Page 31 — Diffraction

1 For noticeable diffraction, the size of the aperture must be roughly equal to the wavelength of the wave passing through it *[1 mark]*. The size of the doorway is roughly equal to the wavelength of sound, so sound waves diffract when they pass through the gap. This allows the person to hear the fire alarm *[1 mark]*. Light has a wavelength much smaller than the size of the doorway, and so diffraction is unnoticeable. This is why the person cannot see the alarm *[1 mark]*.

2 a) E.g. Laser light is monochromatic/only contains one wavelength/frequency of light *[1 mark]*. This provides a clearer pattern than non-monochromatic light sources as different wavelengths diffract by different amounts *[1 mark]*.

b) The central maximum will be wider and less intense *[1 mark]*. Using a narrower slit means more diffraction *[1 mark]* and so there will be fewer photons per unit area, so the intensity will be lower *[1 mark]*.

Page 33 — Two-Source Interference

1 a) Distance between sources (slit width) *[1 mark]*, spacing between two consecutive maxima or minima (fringe spacing) *[1 mark]* and the distance from the sources to the screen *[1 mark]*.

b) Any two from: Don't look into the laser beam / don't point the beam at a person / don't point the laser beam at a reflective surface / display a laser warning sign / wear laser safety goggles / turn the laser off when not in use ***[1 mark for two correct suggestions]***
Laser light can permanently damage your eyes/retinas *[1 mark]*.

2 a) $\lambda = c / f = 330 / 1320 = \mathbf{0.25\ m}$ *[1 mark]*

b) Separation = $w = \lambda D / s = (0.25 \times 7.3) / 1.5 = 1.21666...$
= **1.2 m (to 2 s.f.)** *[1 mark]*

Page 35 — Diffraction Gratings

1 a) Use $d \sin\theta = n\lambda$.
For the first order, $n = 1$, so, $\sin\theta = \lambda / d$ *[1 mark]*
No need to actually work out d. The number of lines per metre is $1/d$. So you can simply multiply the wavelength by that.
$\sin\theta = 6.0\times10^{-7}\times4.0\times10^{5} = 0.24$
$\theta = \sin^{-1}(0.24) = 13.886... = \mathbf{14°}$ **(to 2 s.f.)** *[1 mark]*
For the second order, $n = 2$ and $\sin\theta = 2\lambda / d$.
You already have a value for λ / d. Just double it to get $\sin\theta$ for the second order.
$\sin\theta = 0.48$, $\theta = \sin^{-1}(0.48) = 28.685... = \mathbf{29°}$ **(to 2 s.f.)** *[1 mark]*

b) No. Putting $n = 5$ into the equation gives a value of $\sin\theta$ of 1.2, which is impossible *[1 mark]*.

2 $d \sin\theta = n\lambda$, so for the 1st order maximum, $d \sin\theta = \lambda$
$\sin 14.2° = \lambda \times 3.7 \times 10^{5}$ *[1 mark]*
$\lambda = 6.629...\times10^{-7} = \mathbf{6.6\times10^{-7}\ m}$ **(or 660 nm) (to 2 s.f.)** *[1 mark]*

Page 37 — Refractive Index

1 a) $n_{diamond} = c / c_{diamond} = (3.00\times10^{8}) / (1.24\times10^{8}) = 2.419...$
= **2.42 (to 3 s.f.)** *[1 mark]*

Answers

b) $n_{air} \sin \theta_1 = n_{diamond} \sin \theta_2$, $n_{air} = 1$
So, $\sin \theta_1 = n_{diamond} \sin \theta_2$ ***[1 mark]***
$\sin \theta_2 = \sin 50° / 2.419... = 0.316...$
$\theta_2 = \sin^{-1} (0.316...) = 18.459... =$ **18° (to 2 s.f.)** ***[1 mark]***
You can assume the refractive index of air is 1, and don't forget to write the degree sign in your answer.

2 a) When the light is pointing steeply upwards some of it is refracted and some reflected — the beam emerging from the surface is the refracted part ***[1 mark]***.
However, when the beam hits the surface at more than the critical angle (to the normal to the boundary) refraction does not occur. All the beam is totally internally reflected to light the tank, hence its brightness ***[1 mark]***.

b) The critical angle is 90° – 41.25° = 48.75° ***[1 mark]***.
$\sin \theta_c = n_{air} / n_{water} = 1 / n_{water}$,
so $n_{water} = 1 / \sin \theta_c = 1 / \sin 48.75°$
$= 1.330... =$ **1.330 (to 4 s.f.)** ***[1 mark]***
The question talks about the angle between the light beam and the floor of the aquarium. This angle is 90° minus the incident angle — measured from a normal to the surface of the water.

3 a) The cladding has a lower refractive index than the fibre, to allow total internal reflection ***[1 mark]***. It also protects the cable from scratches and damage which may let light escape ***[1 mark]***.

b) How to mark your answer (pick the level description that best matches your answer):
5-6 marks:
The answer fully describes several potential causes of signal degradation and correctly explains how to alter the design or operation to reduce the effects.
3-4 marks:
One potential cause of signal degradation has been fully described with a full explanation of how to reduce the effect. Another potential cause is mentioned with incomplete information or suggestions for design/operation alterations.
1-2 marks:
Potential reason(s) for signal degradation suggested but with incomplete description of the effects or of suggestions of how to reduce the effects with design changes.
0 marks:
No relevant information is given.
Here are some points your answer may include:
- Absorption of light by the material the fibre is made from causes a loss of signal amplitude.
- Dispersion within the fibre causes pulse broadening.
- Modal dispersion is caused by light rays taking different paths of different lengths down the fibre.
- Material dispersion is caused by different wavelengths of light refracting by different amounts in the fibre.
- Designing the fibre so that it only allows light to take one path through it (known as a single-mode fibre) reduces pulse broadening due to modal dispersion.
- Sending signals using monochromatic light reduces material dispersion because monochromatic light only has one wavelength/frequency.
- Using an optical fibre repeater to boost/regenerate signals can help to reduce signal degradation over long transmission distances.

Section 4 — Mechanics

Page 39 — Scalars and Vectors

1 Start by drawing a diagram:

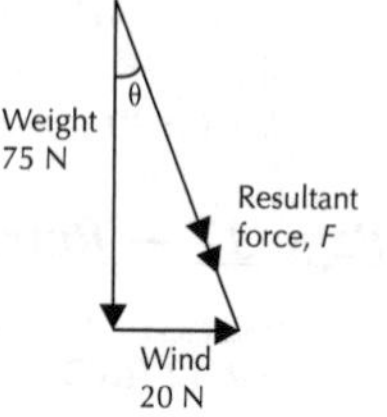

$F^2 = 20^2 + 75^2 = 6025$,
so **F = 78 N (to 2 s.f.)** ***[1 mark]***
$\tan \theta = \frac{20}{75} = 0.266...$
So **θ = 15° (to 2 s.f.) to the vertical** ***[1 mark]***.
Make sure you know which angle you're finding — and label it on your diagram. You could also answer this question by drawing a scale diagram.

2 Again, start by drawing a diagram:
Horizontal component
$v_H = 20 \cos 15°$
= **19 ms⁻¹ (to 2 s.f.)** ***[1 mark]***
Vertical component
$v_V = 20 \sin 15° =$ **5.2 ms⁻¹ (to 2 s.f.) downwards** ***[1 mark]***
Always draw a diagram.

Page 41 — Forces

1 Weight = vertical component of tension × 2
$8 \times 9.81 = 2T \sin 50°$ ***[1 mark]***
$78.48 = 0.766... \times 2T$
$102.45... = 2T$
T = 51 N (to 2 s.f.) ***[1 mark]***

2 By Pythagoras:
$R = \sqrt{1200^2 + 610^2} = 1346.1...$
= 1300 N (to 2 s.f.) ***[1 mark]***
$\tan \theta = \frac{610}{1200}$, so $\theta = \tan^{-1} 0.50...$
= 26.95° = **27° (to 2 s.f.)** ***[1 mark]***

Page 43 — Moments

1 Moment = force × distance
$60 = 0.40F$ ***[1 mark]***, so F = **150 N** ***[1 mark]***

2 Clockwise moment = anticlockwise moment
$W \times 2.0 = T \times 0.3$
[1 mark for either line of working]
$60 \times 9.81 \times 2.0 = T \times 0.3$
$T = 3924 =$ **4000 N (to 1 s.f.)** ***[1 mark]***

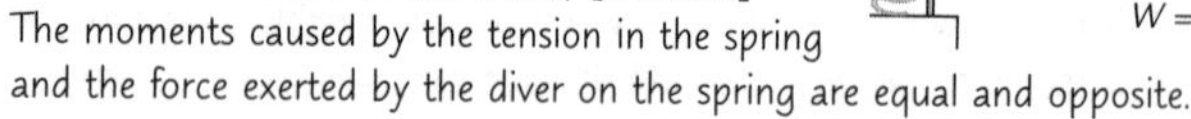

The moments caused by the tension in the spring and the force exerted by the diver on the spring are equal and opposite.

Page 45 — Mass, Weight and Centre of Mass

1 a) Experiment:
Hang the object freely from a point. Hang a plumb bob from the same point, and use it to draw a vertical line down the object ***[1 mark]***. Repeat for a different point and find the point of intersection ***[1 mark]***. The centre of mass is halfway through the thickness of the object (by symmetry) at the point of intersection ***[1 mark]***.
Identifying and reducing error: e.g.
Source: the object and/or plumb line might move slightly while you're drawing the vertical line ***[1 mark]***.
Reduced by: hang the object from a third point to confirm the position of the point of intersection ***[1 mark]***.

b) You can find the centre of mass of any regular shape using symmetry. The centre of mass will be at the centre where the lines of symmetry cross, halfway through its thickness ***[1 mark]***.

Page 47 — Displacement-Time Graphs

1 Split graph into four sections:
A: Acceleration ***[1 mark]***
B: Constant velocity ***[1 mark]***
C: Stationary ***[1 mark]***
D: Constant velocity in opposite direction to A and B ***[1 mark]***

2 a)

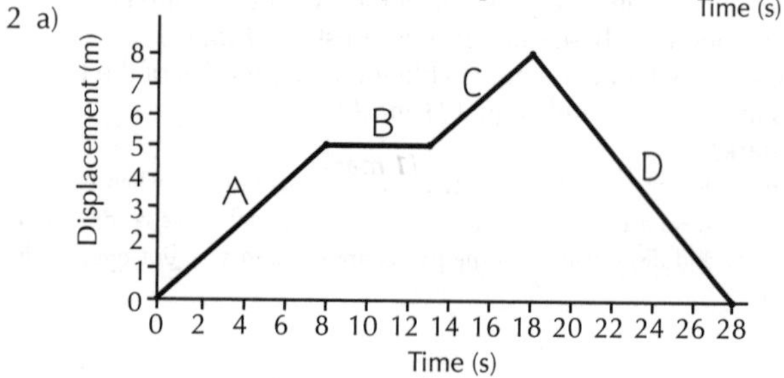

[4 marks available — 1 mark for each section correctly drawn.]

Answers

b) At A: $v = \frac{\text{displacement}}{\text{time}} = \frac{5}{8} = 0.625 = \mathbf{0.6\ ms^{-1}}$ **(to 1 s.f.)**

At B: $v = 0$

At C: $v = \frac{\text{displacement}}{\text{time}} = \frac{3}{5} = \mathbf{0.6\ ms^{-1}}$

At D: $v = \frac{\text{displacement}}{\text{time}} = \frac{-8}{10} = \mathbf{-0.8\ ms^{-1}}$

[2 marks for all correct or just 1 mark for 2 or 3 correct.]

Page 49 — Velocity-Time and Acceleration-Time Graphs

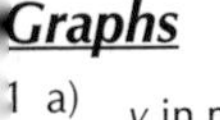

1 a)

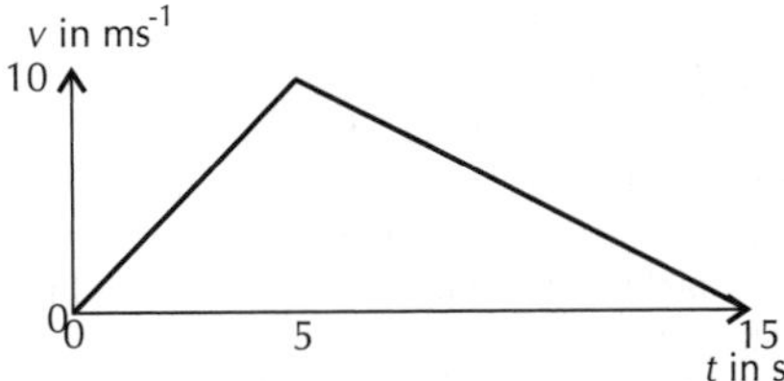

[1 mark for drawing a correctly labelled pair of axes, with a straight line between v = 0 ms⁻¹, t = 0 s and v = 10 ms⁻¹, t = 5 s. 1 mark for correctly drawing a straight line between v = 10 ms⁻¹, t = 5 s and v = 0 ms⁻¹, t = 15 s.]

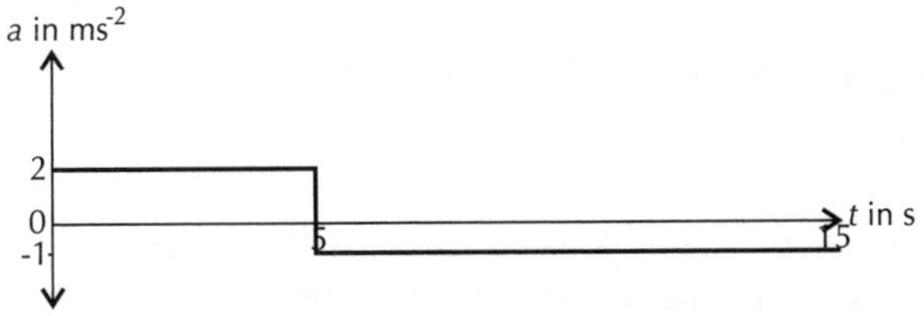

[1 mark for drawing a correctly labelled pair of axes, with a straight line between a = 2 ms⁻², t = 0 s and a = 2 ms⁻², t = 5 s. 1 mark for correctly drawing a straight line from a = 2 ms⁻², t = 5 s and a = –1 ms⁻², t = 5 s, and a straight line between a = –1 ms⁻², t = 5 s and a = –1 ms⁻², t = 15 s.]

b) Distance travelled is equal to the area under v-t graph between $t = 0$ and $t = 5$ ***[1 mark]*** $= 0.5 \times 5 \times 10 =$ **25 m** ***[1 mark]***

Page 51 — Motion With Uniform Acceleration

1 a) $a = -9.81\ ms^{-2}$, $t = 5\ s$, $u = 0\ ms^{-1}$, $v = ?$

Use: $v = u + at$, $v = 0 + 5 \times -9.81$ ***[1 mark]***

$v = \mathbf{-49\ ms^{-1}}$ **(to 2 s.f.)** ***[1 mark]***

NB: It's negative because she's falling downwards and we took upwards as the positive direction.

b) Use: $s = \left(\frac{u+v}{2}\right)t$ or $s = ut + \frac{1}{2}at^2$

$s = \frac{-49}{2} \times 5 = -120$ m (to 2 s.f.)

or $s = 0 + \frac{1}{2} \times -9.81 \times 5^2 = -120$ m (to 2 s.f.)

So she fell **120 m** (**to 2 s.f.**) ***[1 mark for working, 1 mark for answer]***

2 a) $v = 0\ ms^{-1}$, $t = 3.2\ s$, $s = 40\ m$, $u = ?$ Use: $s = \left(\frac{u+v}{2}\right)t$

$40 = 3.2u \div 2$ ***[1 mark]***, so $u = \frac{80}{3.2} = \mathbf{25\ ms^{-1}}$ ***[1 mark]***

b) E.g. use: $v^2 = u^2 + 2as$

$0 = 25^2 + 80a$ ***[1 mark]***

$-80a = 625$, so $a = \mathbf{-7.8\ ms^{-2}}$ **(to 2 s.f.)** ***[1 mark]***

3 a) Take upstream as negative:

$v = 5\ ms^{-1}$, $a = 6\ ms^{-1}$, $s = 1.2\ m$, $u = ?$ Use: $v^2 = u^2 + 2as$

$5^2 = u^2 + 2 \times 6 \times 1.2$ ***[1 mark]***

$u^2 = 25 - 14.4 = 10.6$

$u = -3.255... = \mathbf{-3\ ms^{-1}}$ **(to 1 s.f.)** ***[1 mark]***

Take the negative root as we've defined downstream to be positive and initially the boat was travelling upstream.

b) From furthest point: $u = 0\ ms^{-1}$, $a = 6\ ms^{-2}$, $v = 5\ ms^{-1}$, $s = ?$

Use: $v^2 = u^2 + 2as$

$5^2 = 0 + 2 \times 6 \times s$ ***[1 mark]***

$s = 25 \div 12 = 2.083... =$ **2 m (to 1 s.f.)** ***[1 mark]***

Page 53 — Acceleration Due to Gravity

1 a) The air resistance on a falling small steel ball will be less than that on a beach ball. The air resistance on the ball used in this experiment needs to be negligible in order to be able to calculate the value of g ***[1 mark]***.

b) E.g. the ball's fall might be affected by wind ***[1 mark]***. To remove this error, conduct the experiment indoors and close all windows ***[1 mark]***. / Not aligning the ball and ruler at eye level can lead to a measuring error ***[1 mark]***. This can be reduced by making sure your eye is perpendicular to the measuring scale being used when taking measurements ***[1 mark]***.

c) E.g. there may be a delay on the stopwatch/light gates ***[1 mark]***. To remove this, ensure they are properly calibrated before conducting the experiment ***[1 mark]***. / The ruler may not be aligned properly so would give slightly incorrect vertical height measurements ***[1 mark]***. To remove this, use a clamp to ensure the rule is straight and unmoving ***[1 mark]***.

d) Use: $s = ut + \frac{1}{2}at^2$ ***[1 mark]***

$u = 0$, so $s = \frac{1}{2}at^2$ or $\frac{1}{2}a = \frac{s}{t^2}$ ***[1 mark]***

So the gradient of a graph of s against t^2, $\frac{\Delta s}{\Delta t^2}$, is equal to half the acceleration, i.e. $\frac{1}{2}g$ ***[1 mark]***.

Page 55 — Projectile Motion

1 a) You only need to worry about the vertical motion of the stone.

$u = 0\ ms^{-1}$, $s = -560\ m$, $a = -g = -9.81\ ms^{-2}$, $t = ?$

You need to find t, so use: $s = ut + \frac{1}{2}at^2$

$-560 = 0 + \frac{1}{2} \times -9.81 \times t^2$ ***[1 mark]***

$t = \sqrt{\frac{2 \times (-560)}{-9.81}} = 10.68... =$ **11 s (to 2 s.f.)** ***[1 mark]***

b) You know that in the horizontal direction:

$u = v = 20\ ms^{-1}$, $t = 10.68...\ s$, $a = 0$, $s = ?$

So use velocity $= \frac{\text{distance}}{\text{time}}$, $v = \frac{s}{t}$

$s = v \times t = 20 \times 10.68...$ ***[1 mark]*** = **210 m (to 2 s.f.)** ***[1 mark]***

2 You know that for the arrow's vertical motion (taking upwards as the positive direction):

$a = -9.81\ ms^{-2}$, $u = 30\ ms^{-1}$ and the arrow will be at its highest point just before it starts falling back towards the ground, so $v = 0\ ms^{-1}$.

s = the vertical distance travelled from the arrow's firing point.

So use $v^2 = u^2 + 2as$

$0 = 30^2 + 2 \times -9.81 \times s$ ***[1 mark]***

$900 = 2 \times 9.81s$

$s = \frac{900}{2 \times 9.81} = 45.87... = 45.9$ m ***[1 mark]***

So the maximum distance reached from the ground $= 45.87... + 1 =$ **47 m (to the nearest metre)** ***[1 mark]***

Page 57 — Newton's Laws of Motion

1 a) Force perpendicular to river flow = 500 – 100 = 400 N ***[1 mark]***

Force parallel to river flow = 300 N

Resultant force $= \sqrt{400^2 + 300^2} =$ **500 N** ***[1 mark]***

b) $a = F/m$ (from $F = ma$)

$= 500/250$ ***[1 mark]*** = **2 ms⁻²** ***[1 mark]***

2 B ***[1 mark]***

The overall acceleration is a, so ma must be equal to the resultant force, which is the force John is pushing with minus the resistance caused by friction. So $F_{John} - F = ma$, and $F_{John} = ma + F$.

3 The only force acting on each of them is their weight = mg ***[1 mark]***. Since $F = ma$, this gives $ma = mg$, or $a = g$ ***[1 mark]***. Their acceleration doesn't depend on their mass — it's the same for both of them — so they reach the water at the same time ***[1 mark]***.

Answers

Page 59 — Drag, Lift and Terminal Speed

1 a) The velocity increases at a steady rate, which means the acceleration is constant ***[1 mark]***.
Constant acceleration means there must be no atmospheric resistance (atmospheric resistance would increase with velocity, leading to a decrease in acceleration). So there must be no atmosphere ***[1 mark]***.

b)

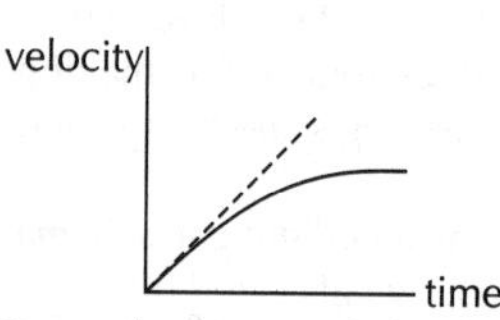

[2 marks — 1 mark for drawing a graph that still starts from the origin, 1 mark for showing the graph curving to show the velocity increasing at a decreasing rate until the velocity is constant.]

Your graph must be a smooth curve which levels out. It must NOT go down at the end.

c) The graph becomes less steep because the acceleration is decreasing ***[1 mark]*** and because air resistance increases with speed ***[1 mark]***. The graph levels out because air resistance has become equal to weight ***[1 mark]***.

If the question says 'explain', you won't get marks for just describing what the graph shows — you have to say <u>why</u> it is that shape.

Page 61 — Momentum and Impulse

1 a) total momentum before collision = total momentum after ***[1 mark]***
$(0.60 \times 5.0) + 0 = (0.60 \times -2.4) + 2v$
$3 + 1.44 = 2v$ ***[1 mark]*** $v = 2.22... =$ **2.2 ms⁻¹ (to 2 s.f.)** ***[1 mark]***

b) Kinetic energy before collision
$= \frac{1}{2} \times 0.6 \times 5^2 + \frac{1}{2} \times 2 \times 0^2 =$ **7.5 J** ***[1 mark]***
Kinetic energy after the collision
$= \frac{1}{2} \times 0.6 \times 2.4^2 + \frac{1}{2} \times 2 \times 2.22^2 = 1.728 + 4.9284$
= 6.7 J (to 2 s.f.) ***[1 mark]***
The kinetic energy of the two balls is greater before the collision than after (i.e. it's not conserved), so the collision must be inelastic ***[1 mark]***.

2 momentum before = momentum after ***[1 mark]***
$(0.7 \times 0.3) + 0 = 1.1v \Rightarrow 0.21 = 1.1v$ ***[1 mark]*** $\Rightarrow$
$v =$ **0.2 ms⁻¹ (to 1 s.f.)** ***[1 mark]***

Page 63 — Work and Power

1 a) Force in direction of travel
$= 100 \cos 40° = 76.6...$ N ***[1 mark]***
$W = Fs = 76.6... \times 1500 =$
110 000 J (to 2 s.f.) ***[1 mark]***

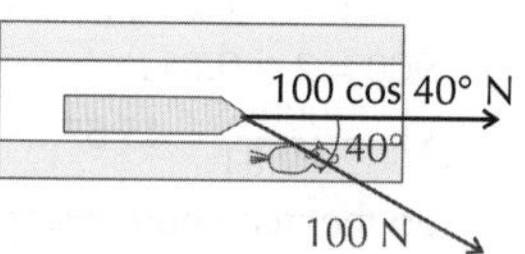

b) Use $P = Fv$
$= 100 \cos 40° \times 0.8$ ***[1 mark]*** = **61 W (to 2 s.f.)** ***[1 mark]***

2 a) Use $W = Fs$
$= 20 \times 9.81 \times 3.0$ ***[1 mark]*** = **590 J (to 2 s.f.)** ***[1 mark]***
Remember that 20 kg is not the force — it's the mass. So you need to multiply it by 9.81 Nkg⁻¹ to get the weight.

b) Use $P = Fv$
$= (20 \times 9.81) \times 0.25$ ***[1 mark]*** = **49 W (to 2 s.f.)** ***[1 mark]***

Page 65 — Conservation of Energy and Efficiency

1 a) Use $E_k = \frac{1}{2}mv^2$ and $\Delta E_p = mg\Delta h$ ***[1 mark]***
$\frac{1}{2}mv^2 = mg\Delta h$
$\frac{1}{2}v^2 = g\Delta h$
$v^2 = 2g\Delta h = 2 \times 9.81 \times 2.0 = 39.24$ ***[1 mark]***
$v =$ **6.3 ms⁻¹ (to 2 s.f.)** ***[1 mark]***
'No friction' allows you to say that the change in kinetic energy is the same as the change in potential energy.

b) 2 m — no friction means the kinetic energy will all change back into potential energy, so he will rise back up to the same height as he started ***[1 mark]***.

c) Put in some more energy by actively 'skating' ***[1 mark]***.

2 a) If there's no air resistance, $E_k = E_p = mg\Delta h$ ***[1 mark]***
$E_k = 0.02 \times 9.81 \times 8.0 =$ **1.6 J (to 2 s.f.)** ***[1 mark]***

b) If the ball rebounds to 6.5 m, it has gravitational potential energy:
$E_p = mg\Delta h = 0.02 \times 9.81 \times 6.5 = 1.28$ J ***[1 mark]***
So $1.57 - 1.28 =$ **0.29 J** is converted to other forms ***[1 mark]***

Section 5 — Materials

Page 67 — Properties of Materials

1 a) Hooke's law says that force is proportional to extension.
The force is 1.5 times as great, so the extension will be 1.5 times the original value.
Extension = 1.5 × 4.0 mm = **6.0 mm** ***[1 mark]***

b) $F = k\Delta L$ so $k = F \div \Delta L$ ***[1 mark]***
$k = 10.0 \div (4.0 \times 10^{-3}) =$ **2500 Nm⁻¹** ***[1 mark]***
You could also use the values for F and ΔL from part a) to work out k.

c) Any from e.g. The string now stretches much further for small increases in force ***[1 mark]***. / When the string is loosened it is longer than at the start ***[1 mark]***.

2 The rubber band does not obey Hooke's law ***[1 mark]*** because when the force is doubled from 2.5 N to 5.0 N, the extension increases by a factor of 2.3 ***[1 mark]***.
You could also work out k for both 2.5 N and 5.0 N, and show that it varies — i.e. the extension is not proportional to the force.

Page 69 — Stress and Strain

1 a) Area $= \pi r^2$ or $\pi\left(\frac{d}{2}\right)^2$

So area $= \pi \times \frac{(1.0 \times 10^{-3})^2}{4} = 7.853... \times 10^7$ ***[1 mark]***
Stress = force/area $= 300 \div (7.853... \times 10^{-7})$
= 3.8 × 10⁸ Nm⁻² (or Pa) (to 2 s.f.) ***[1 mark]***

c) Strain = extension ÷ length
$= (4.0 \times 10^{-3}) \div 2.00 =$ **2.0 × 10⁻³ (to 2 s.f)** ***[1 mark]***

2 a) $F = k\Delta L$ and so rearranging $k = F \div \Delta L$ ***[1 mark]***
$k = 50.0 \div (3.0 \times 10^{-3}) =$ **1.7 × 10⁴ Nm⁻¹ (to 2 s.f.)** ***[1 mark]***

b) Elastic strain energy $= \frac{1}{2}F\Delta L$
$= \frac{1}{2} \times 50.0 \times 3.0 \times 10^{-3}$
= 7.5 × 10⁻² J ***[1 mark]***
You could also use $E = \frac{1}{2}k\Delta L^2$ and substitute in your value of k.

3 The force needed to compress the spring is:
$F = k\Delta L = 40.8 \times 0.05 = 2.04$ N ***[1 mark]***
The elastic strain energy in the spring is then:
$E = \frac{1}{2}F\Delta L = \frac{1}{2} \times 2.04 \times 0.05 = 0.051$ J ***[1 mark]***
Assume all this energy is converted to kinetic energy in the ball.
$E = E_{kinetic} = 0.051$ J ***[1 mark]***.
You could also begin by using Hooke's law to replace F in the formula for elastic strain energy, to give $E = \frac{1}{2}k\Delta L^2$, and then substituting into this.

Page 71 — The Young Modulus

1 Cross-sectional area $= \pi r^2$ or $\pi\left(\frac{d}{2}\right)^2$

So area $= \pi \times \frac{(0.60 \times 10^{-3})^2}{4} = 2.827... \times 10^{-7}$ m² ***[1 mark]***
Stress = force/area $= 80.0 \div (2.827... \times 10^{-7})$
$= 2.829... \times 10^8$ Nm⁻² ***[1 mark]***
Strain = extension/length $= (3.6 \times 10^{-3}) \div 2.50$
$= 1.44 \times 10^{-3}$ ***[1 mark]***
Young modulus = stress/strain
$= (2.829... \times 10^8) \div (1.44 \times 10^{-3})$
= 2.0 × 10¹¹ Nm⁻² (to 2 s.f.) ***[1 mark]***

2 $E = \frac{FL}{\Delta L A}$. Force, original length and extension are the same for both wires, so $E \propto \frac{1}{A}$.
The wire B has half the cross-sectional area of the wire A. So the Young modulus of wire B (E_B) must be twice that of the wire A ***[1 mark]***.
$E_B = 2 \times 7.0 \times 10^{10}$ Nm⁻² $= 1.4 \times 10^{11}$ Nm⁻²
= 1.4 × 10¹¹ Nm⁻² (to 2 s.f.) ***[1 mark]***

Answers

1 a) Young modulus, E = stress/strain and so strain = stress/E
Strain on wire = $(2.6 \times 10^8) \div (1.3 \times 10^{11})$ ***[1 mark]***
= $\mathbf{2.0 \times 10^{-3}}$ **(to 2 s.f.)** ***[1 mark]***

b) Stress = force/area and so area = force/stress
Area = $100 \div (2.6 \times 10^8)$ ***[1 mark]***
= $3.846... \times 10^{-7}$ = $\mathbf{3.8 \times 10^{-7}\ m^2}$ **(to 2 s.f.)** ***[1 mark]***

c) Strain energy per unit volume
= ½ × stress × strain = ½ × $(2.6 \times 10^8) \times (2.0 \times 10^{-3})$ ***[1 mark]***
= $\mathbf{2.6 \times 10^5\ Jm^{-3}}$ ***[1 mark]***
Give the mark if answer is consistent with the value calculated for strain in part a).

Page 73 — Stress-Strain and Force-Extension Graphs

1 a) Liable to break suddenly without deforming plastically ***[1 mark]***.
b) E.g.

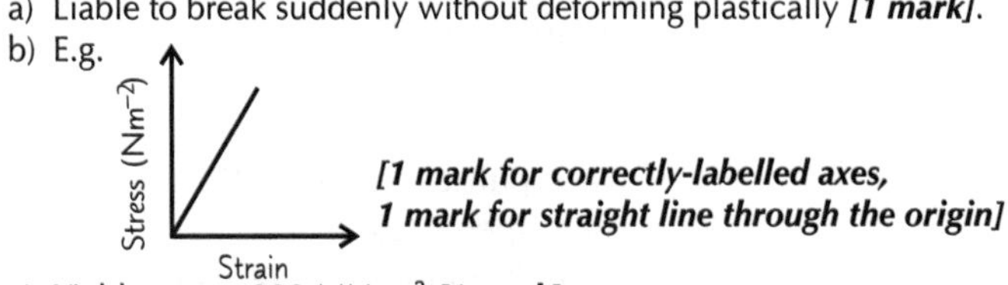

[1 mark for correctly-labelled axes, 1 mark for straight line through the origin]

2 a) Yield stress = 900 MNm^{-2} ***[1 mark]***
b) Energy per unit volume
= area under the graph ***[1 mark]***
= $(800 \times 0.2) \div 2$ = **80 MJ m^{-3}** (or $\mathbf{8 \times 10^7\ J\ m^{-3}}$) ***[1 mark]***
c) Work done to deform the thread per unit volume is the area between the loading curve and the unloading curve ***[1 mark]***.
So work done = area under loading curve – area under unloading curve ***[1 mark]***.

Section 6 — Electricity

Page 75 — Current, Potential Difference and Resistance

1 Time in seconds = 10.0×60 = 600 s. Use the formula $I = \Delta Q / \Delta t$ which gives you I = 4500 / 600 = **7.5 A** ***[1 mark]***
Write down the formula first. Don't forget the unit in your answer.

2 Rearrange the formula $V = W / Q$ to give $Q = W / V$
so you get Q = 120 / 12 ***[1 mark]*** = **10 C** ***[1 mark]***

3 a) $R = V/I$ = 2/12 = 0.166... = **0.17 Ω (to 2 s.f.)** ***[1 mark]***
b) $I = V/R$ = 35/0.166... = **210 A** ***[1 mark]***
c)

210
Current / A
0
0 Potential difference / V 35

[1 mark]

Page 77 — I/V Characteristics

1 a) The graph is curved, starting steep and then levelling off as voltage/current increases ***[1 mark]***.
b) As the current increases, the temperature of the filament increases ***[1 mark]***. As the temperature increases, the resistance increases, so the graph gets shallower ***[1 mark]***.

2 As the temperature of the thermistor increases, more charge carriers are released ***[1 mark]***. More charge carriers available allows more current to flow in the circuit ***[1 mark]***. So the current in the circuit can be used to monitor the temperature ***[1 mark]***.

Page 79 — Resistivity and Superconductivity

1 a) Area = $\pi(d/2)^2$ and $d = 1.0 \times 10^{-3}$ m
so Area = $\pi \times (0.5 \times 10^{-3})^2$
= $7.853... \times 10^{-7}\ m^2$ ***[1 mark]***

$$R = \frac{\rho l}{A} = \frac{2.8 \times 10^{-8} \times 4.00}{7.853... \times 10^{-7}} = \mathbf{0.14\ \Omega\ (to\ 2\ s.f.)}$$

[1 mark for correct working, 1 mark for answer and unit.]

b) Resistance will now be zero ***[1 mark]***.
Because aluminium is a superconductor below its transition temperature of 1.2 K ***[1 mark]***.

2 a) Resisitivity varies with temperature ***[1 mark]*** so she must find the resistivity at 20°C in order to compare it to the resistivities in the table ***[1 mark]***.
b) The diameter / cross-sectional area of the wire ***[1 mark]*** using a micrometer / micrometer caliper / caliper ***[1 mark]***.
c) E.g. the wire is cylindrical / has a circular cross section ***[1 mark]***.

Page 81 — Electrical Energy and Power

1 a) $P = VI$, so heater current = P/V = 920/230 = **4.0 A (to 2 s.f.)** ***[1 mark]***
b) $V = IR$, so motor current = V/R = 230/190
= 1.210... = **1.2 A (to 2 s.f.)** ***[1 mark]***
c) Motor power = $V \times I$ = 230 × 1.210... = 278.4...
= 280 W (to 2 s.f.) ***[1 mark]***
Total power = motor power + heater power
= 278.4... + 920 = 1198 W
= 1200 W (to 2 s.f.) = **1.2 kW (to 2 s.f.)** ***[1 mark]***

2 a) $E = VIt$ = 12 × 48 × 2.0 ***[1 mark]*** = **1200 J (to 2 s.f.)** ***[1 mark]***
b) Energy wasted = I^2Rt = $48^2 \times 0.01 \times 2.0$ ***[1 mark]***
= **46 J (to 2 s.f.)** ***[1 mark]***

Page 83 — E.m.f. and Internal Resistance

1 a) $\varepsilon = I(R + r)$ so $I = \varepsilon/(R + r)$ = 24/(4.0 + 0.80) ***[1 mark]***
= **5.0 A (to 2 s.f.)** ***[1 mark]***
b) $v = Ir$ = 5.0 × 0.80 = **4.0 V (to 2 s.f.)** ***[1 mark]***
You could have used $\varepsilon = V + v$ and calculated V using $V = IR_{drill}$

2 C ***[1 mark]***
$\varepsilon = I(R + r)$, but since there are two cells in series replace r with $2r$, and ε with 2ε, then rearrange to find I.

Page 85 — Conservation of Energy and Charge

1 a) Resistance of parallel resistors:
$1/R_{parallel} = 1/6.0 + 1/3.0 = 1/2 \Rightarrow R_{parallel} = 2.0\ \Omega$ ***[1 mark]***
Total resistance:
$R_{total} = 4.0 + R_{parallel} = 4.0 + 2.0 =$ **6.0 Ω** ***[1 mark]***
b) $V = I_3R_{total} \Rightarrow I_3 = V / R_{total}$ = 12 / 6.0 = **2.0 A** ***[1 mark]***
c) $V = IR$ = 2.0 × 4.0 = **8.0 V** ***[1 mark]***
d) E.m.f. = sum of p.d.s in circuit, so 12 = 8.0 + $V_{parallel}$
$V_{parallel}$ = 12 – 8.0 = **4.0 V** ***[1 mark]***
e) $I = V/R$, so I_1 = 4.0 / 3.0 = **1.3 A (to 2 s.f.)** ***[1 mark]***
I_2 = 4.0 / 6.0 = **0.67 A (to 2 s.f.)** ***[1 mark]***
You can check your answers by making sure that $I_3 = I_2 + I_1$.

Page 87 — The Potential Divider

1 Parallel circuit, so p.d. across both sets of resistors is 12 V.
a) There are two equal resistors in the top branch of the circuit. The p.d. between points A and B is equal to the potential difference across one of these resistors:
$V_{AB} = \frac{1}{2} \times 12 = 6.0$ V ***[1 mark]***
b) There are three equal resistors in the bottom branch of the circuit. The p.d. between points A and C is equal to the potential difference across two of them:
$V_{AC} = \frac{2}{3} \times 12 = 8.0$ V ***[1 mark]***
c) $V_{BC} = V_{AC} - V_{AB} = 8 - 6 = 2.0$ V ***[1 mark]***

2 a) V_{AB} = 50/80 × 12 = 7.5 V ***[1 mark]***
(ignore the 10 Ω — no current flows that way)
b) Total resistance of the parallel circuit:
$1/R_T = 1/50 + 1/(10 + 40) = 1/25 \Rightarrow R_T = 25\ \Omega$ ***[1 mark]***
p.d. over the whole parallel arrangement = 25/55 × 12 = 5.45... V ***[1 mark]***
p.d. across AB = 40/50 × 5.45...
= 4.36... V = **4.4 V (to 2 s.f.)** ***[1 mark]***
current through 40.0 Ω resistor = V/R
= 4.36.../40.0 = **0.11 A (to 2 s.f.)** ***[1 mark]***

Index

A

absolute uncertainties 90
absorption (of a signal) 37
acceleration 46
  due to gravity 52-55
  from velocity-time graphs 48
  Newton's 2nd law 56
  uniform 50, 51, 54, 55
acceleration-time graphs 49
accuracy 93
air bags 61
air resistance 58, 59
alpha emission 4
ammeters 74
amplitude (of a wave) 22
angle of incidence 36
angle of refraction 36
annihilation 7
antibaryons 10
antimatter 6, 7
antineutrinos 5, 6, 9, 11
antineutrons 6, 10
antinodes 28
antiparticles 6, 7
antiprotons 6, 10
antiquarks 13-15
atomic number 2
atoms 2

B

baryon number 10, 11, 13, 14
baryons 10, 11
  quark composition 13
batteries 82
beta-minus decay 5, 9, 11, 14
beta-plus decay 9, 14
breaking stress 68
brittle materials 72, 73

C

calibration 89
car safety features 61
categoric data 92
centre of mass 44, 45
ceramics 72
charge 74
  carriers 76, 77
  conservation 84
  relative 2
  specific 3
circuits 74-87
  batteries 82
  charge 74, 84
  current 74
  diodes 77
  electrical energy 81, 84
  electromotive force (e.m.f.) 82-84
  filament lamps 76
  internal resistance 82, 83
  I/V characteristics 76, 77
  light-dependent resistors (LDRs) 86
  Ohmic conductors 75, 76
  parallel 84
  potential difference 74
  potential dividers 86, 87
  power 80
  resistance 74-78
  series 84
  thermistors 77, 86
cladding (optical fibres) 36
coherence 27
collisions 60, 61
combining uncertainties 90
components of a vector 39, 62
conclusions 93
conservation
  in particle interactions 13, 14
  of charge 84
  of energy 60, 64, 65
  of momentum 60
constructive interference 26, 27
continuous data 92
correlation 92
couples 43
critical angle 36
crumple zones 61
current 74, 84
cycle (of a wave) 22

D

data
  drawing conclusions 93
  evaluation 93
  trends 92
  types 92
data-loggers 49
de Broglie wavelength 20, 21
density 66
dependent variables 88
destructive interference 26, 27
diffraction 30, 31
  gratings 34, 35
  of electrons 20
  patterns 20, 21, 30
  Young's double slit 32, 33
diodes 77
discrete data 92
dispersion (of a signal) 37
displacement 38, 46-48
  of a wave 22
displacement-time graphs 47-49
drag 58
driving force 58

E

efficiency 64
elastic collisions 60
elastic deformation 67, 72, 73
elastic limit 66, 68, 72, 73
elastic potential energy 64
elastic strain energy 64, 67-69, 71
electromagnetic forces 8, 9
electromagnetic radiation 6, 22, 24
  diffraction 30-35
  photons 6, 16
  polarisation 24, 25
  refraction 22, 36
  spectrum 6
  wave-particle duality 20, 21
  wave speed 23
  Young's double slit 32, 33
electromotive force (e.m.f.) 82-84
electrons 2
  current 74
  diffraction 20
  fluorescent tubes 18
  in atoms 2, 18
  in beta-minus decay 5, 11
  pair production 7
  photoelectrons 16, 17
  properties 2, 6, 12
electronvolts 18
electrostatic force between nucleons 4
energy 6
  carried by a photon 6, 16
  conservation 60, 64, 65, 84
  elastic strain energy 64, 67-69, 71
  elastic potential energy 64
  energy levels of atoms 18, 19
  rest energies 6, 7
  transfer 62-65
  transfer (in circuits) 80-82, 84
equations of motion 50, 51, 54, 55
equilibrium 40
error bars 91
ethical experiments 88
exchange particles 8
excitation of electrons 18

F

falling 56, 59
Feynman diagrams 8, 9
filament lamps 76
first harmonic 28
fluorescent tubes 18
force-extension graphs 66, 68, 73
forces
  compressive 66
  couples 43
  electromagnetic 8, 10, 11
  electrostatic 4

Index

exchange particles 8
free-body diagrams 40
friction 58, 59, 64
fundamental forces 8
moments 42, 43
Newton's Laws 56, 57, 61
resolving 40, 41
strong nuclear 4, 8, 10
tensile 66
weak 8, 14
within nuclei 4
orward bias 77
ractional uncertainties 90
racture 72, 73
ree-body force diagrams 40
requency 22, 23
riction 58, 59, 64
ringe spacing 33
undamental forces 8
undamental particles 12, 13

G

gauge bosons 8, 9
graphs
error bars 91
force-extension 66, 68, 73
I/V characteristics 76, 77
line graphs 92
scatter plots 92
stress-strain 68, 71-73
gravitational fields 44
gravitational potential energy 64
gravity 8, 44, 52-54, 56

H

hadrons 10, 11
harmonics 28
Hooke's law 66, 72

I

impulse 61
independent variables 88
inelastic collisions 60
intensity of light 31
interference
constructive 26, 27
destructive 26, 27
diffraction patterns 30, 31, 34, 35
of microwaves 32
two-source interference 32, 33
internal resistance 82, 83
ionisation 18
isotopes 3
I/V characteristics 76, 77

J

joule (definition of) 62

K

kaons 11, 14
kinetic energy 64
conservation 60, 64, 65
photoelectric effect 16, 17
Kirchhoff's laws 84, 85

L

lasers 32
lepton number 12, 14
leptons 12
levers 42
lift 58
light
diffraction 30, 31
diffraction gratings 34, 35
electromagnetic spectrum 6
intensity 31
interference 26, 27, 32, 33
photons 6, 16
polarisation 24, 25
reflection 22
refraction 22, 36
signals 36, 37
wave-particle duality 20-21
light-dependent resistors (LDRs) 86
light emitting diodes (LEDs) 77
light sensors 86
limit of proportionality 66, 72, 73
line graphs 92
line of best fit 91
line spectra 18, 19
longitudinal waves 24

M

mass 44
centre of 44, 45
number 2
relative atomic mass 2
material dispersion 37
matter 6, 7
mesons 10, 11, 13, 14
microwaves 28, 32
modal dispersion 37
moments 42, 43
momentum 20, 60, 61
monochromatic light 30
Mr T 79
muons 12

N

neutrinos 5, 6, 12
neutrons 2, 3, 6, 10
nuclear decay 5, 11, 14
quark composition 13
Newton's 1st law 56
Newton's 2nd law 56, 61
Newton's 3rd law 57
nodes 28
NTC thermistors 77, 86
nuclear model of the atom 2
nuclear separation 4
nuclei 2, 4
forces within nuclei 4
nucleon number 2
nucleons 2, 4
nuclide notation 2

O

ohm (definition of) 74
ohmic conductors 75, 76
Ohm's law 75
optical fibres 36, 37
ordered (ordinal) data 92

P

pair production 6, 7, 15
parachutes 59
parallel circuits 84
multiple cells in parallel 82, 83
particle exchange 8
path difference 27
peer review 21
percentage uncertainties 90
period 22
phase difference 22, 26, 27
phase (of a wave) 22, 26, 27
photoelectric effect 16, 17, 20
photoelectric equation 17
photoelectrons 16
maximum kinetic energy 17
photons 6-8, 16-19
pions 11, 14
plastic deformation 67, 72, 73
polarisation 24, 25
potential difference 74
electromotive force (e.m.f.) 82-84
terminal 82
potential dividers 86, 87
potentiometers 87
power
electrical 80
mechanical 63
precision 93
progressive waves 22, 23

Index

proton number 2
protons 2, 6, 10
  nuclear decay 5, 11, 14
  quark composition 13
pulse broadening 37

Q

quark confinement 15
quarks 13-15

R

radioactive isotopes 3
random errors 89
reaction force 54, 57
reflection 22
refraction 22, 36
refractive index 36
repeatability 93
resistance 74-78
  internal resistance 82, 83
  I/V characteristics 76, 77
resistivity 78, 79
resolving vectors 39-41
resonant frequencies 28, 29
resultant forces 41, 56, 58
reverse bias 77
risk assessments 88

S

Santa 28
scatter plots 92
seat belts 61
semiconductors 76
series circuits 84
  multiple cells in series 82
signal degradation 37
Snell's law 36
sound waves 22, 24, 27, 28, 30
specific charge 3
speed
  of waves 23
  terminal 58, 59
springs 66
stability of objects 45
stationary waves 28, 29
  microwaves 28
stiffness constant 66
stopping potential 17
strain (tensile) 68, 70, 71
strangeness 13, 14
stress (tensile) 68, 70, 71
stress-strain graphs 68, 71-73
strong nuclear force 4, 8, 10
superconductors 79
superposition 26, 27
symmetry of cuteness 44
systematic errors 89

T

television signals 25
temperature sensors 86
tensile strain 68, 70, 71
tensile stress 68, 70, 71
terminal potential difference 82
terminal speed 58, 59
thermistors 77, 86
threshold frequency 16, 17
total internal reflection 36
transverse waves 24
two-source interference 32, 33

U

ultimate tensile stress 68
uncertainties 90, 91
  absolute 90
  combining 90
  error bars 91
  fractional 90
  percentage 90
  significant figures 91

V

valid results 93
validating theories 21
vector quantities 38, 39
velocity 46-51, 54-56, 60, 63
velocity-time graphs 48
V/I characteristics 76, 77
virtual particles 8
viscosity 58
voltage 74
voltmeters 74
volume 66

W

watt (definition of) 63
wave cycle 22
wave equation 23
wave-particle duality 20, 21
wavelength 22, 23
waves
  absorption 37
  coherence 27
  diffraction 30, 31
  diffraction gratings 34, 35
  dispersion 37
  frequency 22
  interference 26, 27
  polarisation 24, 25
  progressive waves 22, 23
  reflection 22
  refraction 22, 36
  speed 23
  stationary waves 28, 29
  superposition 26, 27
  wave theory of light 16, 20
  Young's double slit 32, 33
W bosons 8, 9
weak interaction 8, 9, 11, 12, 14
weight 44
white light 30, 35
work 62, 63
work function 16, 17

X

X-ray crystallography 35

Y

yield point 72
Young modulus 70-72
Young's double-slit 32, 33

PAR53